The One Culture

Pergamon Unified
Engineering Series

Pergamon Unified
Engineering Series

Theme:

Importance of History of Sci/Tech as bridge between the two cultures

The One Culture

By

William H. Davenport

Harvey Mudd College
Claremont, California

Pergamon Press

New York/Toronto/Oxford
Sydney/Braunschweig

PERGAMON PRESS INC.
Maxwell House, Fairview Park, Elmsford, N.Y. 10523

PERGAMON OF CANADA LTD.
207 Queen's Quay West, Toronto 117, Ontario

PERGAMON PRESS LTD.
Headington Hill Hall, Oxford

PERGAMON PRESS (AUST.) PTY. LTD.
Rushcutters Bay, Sydney, N.S.W.

VIEWEG & SOHN GmbH
Burgplatz 1, Braunschweig

Library of Congress Catalog Card No. 70-106054

Printed in the United States of America

08-016322-X

Pergamon Unified Engineering Series

For Isobel

Contents

Preface

THIS BOOK is designed to be of use to any student or layman who is interested in or concerned about the state of culture, particularly American culture, in this Age of Technology. It is not an attempt to match the style of C.P. Snow (Lord Snow), the acerb retorts of F.R. Leavis, or the aplomb of Lionel Trilling, the first being the celebrated scientist and novelist who wrote *The Two Cultures* and the latter pair perhaps his most prominent critics. Nor is it an attempt to set up a uniform, standard culture in which assembly-line automata march off the job daily when the whistle blows in order to get home for the TV news. Rather, it is an attempt to review the two culture argument ten years afterwards, to document the continuing battle between technology and the humanities, to bring Snow up to date (there having been vast changes in both British and American educational systems since he wrote), and to appeal for a broader culture in which a man may specialize still, but lead a fuller life and be of more value to society. I have long felt that there was an argument for bringing the cultures together. In the following pages the reader may discover what it is. I can count on at least one strong supporter—Paul Goodman, who has written the following in *The New York Review:*

> In *The Two Cultures,* C.P. Snow berated the humanists for their irrelevance when two-thirds of mankind are starving and what is needed is science and technology. They have perhaps been irrelevant; but unless technology is itself more humanistic and philosophical, it is of no use. *There is only one culture.* (italics mine)

In a way this book is an anthology of ideas with a running connecting commentary by myself, an eclectic synthesis if you will. In addition to my obvious debt to the sources quoted in the following, I owe much in a personal way to many people and I should like to thank them for their help, advice, and criticism. At Harvey Mudd College, my home base since 1957, I am grateful to Joseph Platt, president; William Radley, John Rae, George McKelvey, Zaner Faust, the college research committee, and Margaret Thompson for various forms of assistance and encouragement. Going back to the early sixties, I must express appreciation to the many friends and colleagues at U.C.L.A. who almost down to the present worked with me on the Educational Development Program in the School of Engineering supported by the Ford Foundation. As consultant to the Humanities Sub-committee, I learned a good deal about how to blend humanities, social sciences, and engineering in general courses. Much of the material in Chapter III reflects this program. In particular I thank the late Dean L.M.K. Boelter and Professor Allen Rosenstein, co-principal investigators, Bonham Campbell, Jacob Frankel (now faculty Dean at Harvey Mudd), and Daniel Rosenthal, my collaborator on a Pergamon text named *Engineering: Its Role and Function in Human Society,* and now a part-time colleague in Claremont, working on the Sloan Foundation with me on the continuing problem of blending engineering and humanities.

In the summer of 1968, I spent a month in England and Scotland visiting nine new technological universities at the suggestion of Sir Eric Ashby, Vice-Chancellor of Cambridge University. These institutions, which were not around when Snow wrote, are incorporating and planning courses in social sciences and humanities as part of a slowly developing program in British education which approaches the American ideal of learning for all who can and will study. Here, from among the many who welcomed and helped me, I salute the following: Vice-Chancellor Topping, at Brunel; Sir Robert Burley, at City; A.M. Duncan of Loughborough; Donald Cardwell, of Manchester Institute of Science and Technology; Keith Reader, at Rugby College of Engineering Technology; D.R. Gordon, Strathclyde; and Vice-Chancellor Moore and Professor Gerald Walters, at Bath. My visits, which are discussed also in Chapter III, were made possible by grants from Harvey Mudd College and Pergamon Press, the staff of the latter having been most courteous and cooperative on both sides of the Atlantic. I am particularly indebted to Robert Maxwell, M.P., Detlev Raymond, Frederic Squires, Henry Paasonen, Sylvia M. Halpern and her staff.

I passed the academic year 1968-69 on sabbatical at Harvard University where, thanks to the invitation of Emmanuel Mesthene, Director of the Program on Technology and Society, I did research, attended seminars, and met and talked with many helpful individuals. Professor Everett Mendelsohn of the Department of History of Science at Harvard had made me a guest of the Department, where I also profited from talks with its chairman, I. Bernard Cohen. The Program made it possible to attend the frequent lunches of the Science and Public Policy group as my education continued. While I was working at Widener Library and at 61 Kirkland Street, Cambridge, my life was made fuller and more enjoyable by many people, far too many to thank properly. At the Program, where most of Chapter II was born, I owed much to Tom Parmenter, Charles Hampden-Turner, Mitzi Gerrish, and Jane Draper and her staff, particularly Sheryl Haines, Kat Wright, and Patti Gordon.

While at Harvard, I was most fortunate to visit, talk with, and listen to such established writers and scholars as Don Price, Gerald Holton, and Lewis Mumford, all of whom went out of their way to be helpful. At nearby Simmons College, Professor Wylie Sypher, author of the recent *Literature and Technology,* was an indefatigable correspondent as was Professor Leo Marx at not-too-distant Amherst. Others who were kind enough to answer questions by mail include Joseph Wood Krutch, J. Bronowski, Archibald MacLeish, Martin Esslin, and Leonard Pronko. This is not mere name dropping. The kindness of all these ladies and gentlemen was far beyond the call of duty. I wish I could do more than merely thank them.

There are many others who have directly and indirectly contributed to the making of this book, and I am sure to have failed to mention all the proper names. To them my apologies for inadvertent omission. Finally, to my wife Isobel, I owe more than I can say for keeping me going through the coldest and snowiest Cambridge winter since '88. Without her help I never could have finished the job. And at this writing during the last days of a long hot California summer, I appreciate the morale maintenance further provided by my daughter Linda. There is now nothing more to say than to invoke the ancient formula, "Go, little book," and hope it will provide some teaching and some delight.

Claremont, California — WILLIAM H. DAVENPORT

Chapter I

The Two Cultures: Another Look at Images and Attitudes

"They have a curious distorted image of each other."
–C.P. Snow

"To everyone's relief the squabble about the two cultures has subsided," wrote an astute art critic quite recently in his introduction to a work on literature and technology, adding later on a reference to "the wearisome debate about our two cultures."[1] Wearisome the debate may be, but it has definitely not subsided. The phrase itself is here to stay for a long time. In fact, its occult force is comparable to that of "Strength Through Joy" or "The Great Society." It has a ring. In the tenth year since C.P. Snow delivered his famous Rede lecture and published it (1959), the Dean of Harvard College, Franklin L. Ford, offered a program for science within the humanities in an effort to stop excessive specialization which "threatens to splinter our general culture into not two cultures (*pace* C.P. Snow), but an infinity of narrowly circumscribed cultures having nothing to say to one another."[2] In the previous June, merely to cite one more example, S.K. Overbeck had written the following in a newsletter on art and technology:

> Several years ago, discussions of art and technology began with the handy citation of C.P. Snow's famous 'two cultures' assertion that art and science did not mix, that 'literary culture,' traditionally the lodestone of the arts, was separated from 'scientific culture' by a yawning gap. In the most modern arts today, the gap no longer yawns. It hardly exists.[3]

Two cultures may be trying to become one, or one and a half, or three, as we shall soon see, but there is little doubt that the phrase is still very much in the minds and conversation of academics, intellectuals, and informed laymen.

Writing eighteen years earlier than Snow, Herbert J. Muller noted that science and literature have been at odds since the nineteenth century and that, while there can be no absolute antagonism between such major interests, "the old combatants still sniff at one another; they still want victory."[4] Gerald Walters explains what happened to the original unitary culture that broke down:

> The new industrial middle class found social respectability in an educational system which perpetuated, in a revised form, the traditional classical attitudes, in which there was little room for science and none for technology. Pure science ultimately recovered social status by the end of the nineteenth century, but not, if the Snow thesis is to be accepted, to the point where it became an integral part of a living culture.[5]

It remained for C.P. Snow to tell university people what many of them suspected and some knew for a fact. The main points of his original lecture and the sequel four years later *(A Second Look)* may need refreshing for some readers and are probably unknown to others; accordingly, at this point I shall do a quick summary. Snow felt that more and more polarization was taking place between literary intellectuals and scientists.[6] The scientists are ignorant of Shakespeare and the humanists of the Second Law of Thermodynamics. This is a great loss for us all, and we had better rethink our education, Snow went on, for "three menaces . . . stand in our way–H-Bomb war, overpopulation, the gap between the rich and the poor. This is one of the situations where the worst crime is innocence."[7] Actually, of course, Snow's lecture "was a plea for more knowledge of science among literary intellectuals and more knowledge of literature among scientists," as the *Spectator* editorialized later.[8]

At the time, however, all was not so straightforward and simple. Snow had struck several nerves, and it quickly became apparent in the press on both sides of the Atlantic that the topic of two cultures was one on which many were prepared (and unprepared) to speak. The loudest voice in England was that of F.R. Leavis, whose manner was offensive to many because he turned his criticism of Snow into a personal attack. A representative voice, more measured in tone, in America was that of Lionel Trilling, who, in a sense, put a pox on both their houses and took a course of his own.[9]

Snow has often been misreported and misrepresented. In his sequel he sees some hope of a third culture composed of those concerned with the human effects of the scientific revolution. Yet at the time of the original lecture some, ironically enough, thought this scientist-novelist was pleading *for* the maintenance of two cultures! As he said to me in an interview in the summer of 1968, "Good God, can't they read?" Snow may have his blind spots, underplaying politics and downgrading the authority of literature. Leavis finds his morality weak, and others, in an age of rampant technology, have had fun sniping at his expense.

The tone of various reactions to the science-humanities debate varies greatly according to the background and intelligence of the particular writer, his temperament, and his attitude toward life in general. Thus a philosopher may view both cultures as transient phenomena in the evolution of culture, or as just two of many factors. The biologist—René Dubos is one good example—studies scientific method as one more historical adaptation to environment. If there is a continuum from physics to sociology, as Planck says, a philosopher like W.T. Jones can see a similar continuum or an extension of the same one via semantics, thought, and vision.[10] Even old Matthew Arnold, for all his friendly argument with T.H. Huxley in the last century, wrote that a genuine humanism was scientific. (It must be pointed out that this does not mean that humanities are science.) Even in argument, Huxley quoted Shakespeare and the Bible for support.[11] Today, a Conant writing in the context of modern science and modern man will use metaphor in the manner of a poet. J. Bronowski is both physicist and poet, literally, and calls for scientific humanism, as we shall see later on. And, again, it was the scientist Huxley who wrote that he would be sorry indeed to see a college of science turn out none but lopsided men. What, then, has all the fuss been about? Why dig up the subject again? Don't leading educators today try to humanize engineers and technologize humanists? The answer to the last question is "yes," but the education picture is so complex that I shall devote Chapter III to it. The reason for all the fuss is this: The debate continues; there are some rich and varied comments that any intelligent person should be acquainted with, and in this age of involvement there are definite reasons for responsible people to know why it is crucially necessary for any gulf between cultures to be bridged. The last point will be discussed further in Chapter IV. For the moment let us examine some attitudes toward the two cultures coming out of Snow's book, then go back and look at the images that exist of scientists and humanists and engineers and try to see why and how they do. I will trace my own background and history as one case study that may be, unfortunately, typical. Even the general public, worried about what they read in the paper about man versus Machine, the Bomb, and the so-called "plight of the humanities," cannot stay on the sidelines and say that none of this fuss concerns them.

Whatever may be said about Snow, he started something, and it is an interesting study by itself to see not only how his magic phrase persists but what, in varying ways, intelligent observers have been doing with and to it. Northrop Frye takes a dim view, for example, of any gulf-bridging, but goes further to point out that we have all missed the main issue involving science and humanism:

> Again, the intellectual separation of the 'two cultures' is said to be a problem of our time, but this separation is inevitable; it is going steadily to increase, not decrease, and it cannot possibly be cured by having humanists read more popular science or scientists read more poetry. The real problem is not the humanist's ignorance of science or vice versa, but the ignorance of both humanist and scientist about the society of which they are both citizens.[12]

Where Frye feels that two cultures have both missed contact with the supreme reality, his fellow Canadian, J. Tuzo Wilson, an eminent geophysicist, thinks in terms of a three-culture society made up of arts, humanities, and science, and that the new scientist should blend them.[13] Wylie Sypher, on the other hand, has a different threesome line-up; he sees the battle as between technology on one side and science and art on the other, and feels, furthermore, that engineering is an enemy of art, the latter an unfortunate viewpoint in the light of ancient history and the most modern concept of engineer and sculptor working as a team. However, this literary and articulate academic, writing in late 1968, redeems himself later on, as we shall see, in a perceptive study of art's ethic of waste versus technology's ethic of thrift, with some surprising conclusions. [14] Perhaps we will yet talk of the two ethics!

If critics agree that we have moved away from a unitary culture but disagree on the importance or character of the two or more cultures, we still realize that while we may not be Snow-bound, to use Overbeck's phrase, we are not Snow-free. A third look at his *Two Cultures* reminds us, of course, that in 1959 he did not know much about humanities education in American technical schools; he confined his remarks pretty much to the English scene and the limitations of the class system, and he did show a curious myopia on occasion when he made statements like "There seems to be no place where the cultures meet."[15] One need look no farther than a modern aircraft plant to find cultures meeting on a practical level. For example, in a large plane manufacturing company in Southern California test pilots were complaining that their "butts hurt" after returning from a trial flight; it was finally discovered, after an anthropologist familiar with the history of development of the human frame had been called in to investigate, that pilots' rear ends had grown two inches in width since World War II and the seats of the test planes had been built to earlier specifications. Again, in the early days of space vehicles, those in charge had some difficulty with temperamental displays on the part of fearless astronauts who were willing to brave the unknown but would have no part of some of the known, namely, eating baby food from tubes (too suggestive of infantilism) or submitting to high-colonic irrigation before launching (invasion of privacy). In the latter illustration, a psychologist was called in and eventually the men won both their points. It would have been easy to order submission to these procedures, but the overall feeling on the part of the consultant was that the men might store a bit of resentment away in their minds which could blur or suspend a split-second decision under pressure, and it simply wasn't worth the chance. In all fairness to Snow, however, we must note the word "seems" in his statement; it could mean apparently (but not really) or apparently (obviously so).

The passing of time has, perhaps naturally, removed some of the effect of certain Snow generalizations. Readers will recall his finding it bizarre that so little modern science had been assimilated into modern art. Recent technology, however, which is scientific discovery plus tools and expertise, knows differently, as anyone can testify who attended the exhibition of art and the Machine at the Museum of Modern Art in New York during the winter of 1969. The new note was sounded clearly by Douglas Davis in *Art in America,* and with specific reference to the novelist: "Snow made much of the fact that 'the literary culture' had little of practical use to give 'the scientific culture'. . . . Now every new sign indicates the relationship is a good deal more complicated. . . .When art and science interact, clearly, the new tools can be extended by both."[16] Davis goes on to discuss new materials and forms, new tools and methods, new imagery suggested by tangible technological forms, in short, a new full partnership between artist and Machine in the creative process. More will be said about this in the next chapter.

Even Snow's dichotomy between scientist and non-scientist is not thought to be the most fundamental one by Cyril Bibby, who finds him relevant but not very novel. Writing on the subject "Science: Tool of Culture," Bibby identifies the two cultures as consisting of scientists and artists on the one hand, both exploring the patterns of nature, and "purely verbal scholars" on the other.[17] As Bibby sees it, science is a triumph of both imagination and intellect and is both beautiful and useful; if he is correct, it remains only for the non-artist scholars to come out of their towers and restore the ancient unitary culture. This, needless to say, will take some doing. Another who will have little part of the Snow thesis is Professor Lynn White of U.C.L.A., former college president and outstanding historian of science. In the year of Snow's *Second Look*, White said the following in the course of an address at Cuernevaca in February:

> The move toward the center; the recognition that science and technology are integral to mankind's adventure; the building of a democratic culture: these are clearly only three aspects of one thing which is occurring—the achievement of a unity of human knowledge and experience such as no earlier age has ever conceived. I decline to bewail intellectual disintegration, the schism of Two Cultures, gathering gloom, the stagnation of excess entropy. The evidence of my eyes and ears shows me a tide of the mind which is the exact reverse.[18]

I do not wish to fire off all my powder too early in the skirmish. However, at this point let it be said that White's words sum up beautifully the wider view I am working up to in my last chapter. All of us, and there have been many in the books and articles on the two cultures since 1959, whether for, against, or neutral with regard to Lord Snow, must confess a tremendous debt to this writer, scientist, and gentleman, for making us concentrate on one of the great questions of our time, for timing his remarks with professional artistry, and for expressing his opinions with a clarity without which much intellectual and practical progress now evident might not have materialized.

Many observers would not share Lynn White's optimism or my hopes. One of these is Gerald Holton, professor of physics at Harvard and founder of the intellectual quarterly *Daedalus.* Nine years ago, in what today would be called a "seminal" article, one which anyone interested in the two cultures issue should certainly read, Holton discussed the disturbing implications of the dissociation of science from the rest of our culture, the oft-expressed hope that science and general ideas might work in harness together, and the further hope that the gap between scientists and humanists might be bridged.[19] His flat conclusion was chilling: "But the truth is that both the hopes and the bridges are illusory . . . there appears at present to be no force in our cultural dynamics strong enough to change this trend." In 1969, at this writing, Professor Holton sees no reason to change his mind. No longer is it merely a question of communication between two cultures; it is, because of growing specialization, a matter of communication gap between members of even one scientific department, such as physics.[20] As if this were not enough, one encounters warnings about the difficulties in joining forces by reading the well-known archaeologist, Jacquetta Hawkes, who feels that the imagination required by people to get together is slowly being drained out of modern man along with all his intuitive imagination because of our growing and continuing separation from the land and the tendency of assembly-line technology to turn us all into human automata:

> Years of intellectual or technical training and living can cut off a man's imaginative roots. And from this condition he can only be redeemed by passionate love, by drugs, or other stirrers-up of the psychic depths. (This is a fact that would-be reconcilers of the Two Cultures tend to underestimate.)[21]

The Hawkes quotation about redemption suggests Auden's famous line, "We must love one another or die." As a "would-be reconciler," I would dare edit the line in the present context to read, "We must understand each other or die." And because I feel the effort must be made, I plan to explore some of the problems of communication, the vexed subject of false images and stereotypes, the unhappy business of blind and selfish attitudes, and also the layman's stake in all this, in an attempt to reach such understanding. This material may be considered as a gloss of Snow's original text, brought up to date; it will occupy the concluding portion of this chapter. The second chapter will consider in detail a typical modern confrontation between humanist, artist, and layman on one side, and scientist, engineer, and technologist on the other, in terms of the interplay between the two sides in literature and art since Hiroshima, noting not only the presence of the Bomb but also the continuing mixed love affair and conflict between modern man and the Machine. In the third chapter I should like to show a typical modern attempt to bridge the so-called gap via education, as seen in studies of humanities and fine arts curricula in schools of engineering and technology here and abroad. Finally, as stated above, the fourth chapter will tackle the perhaps impossible task of showing why a restoration of one general culture is imperative and how in many approaches it could be done if enough human beings are interested to the point of making an effort.

Before this begins to sound too sticky or evangelical, let's go back and analyze some of the reasons why we apparently got into our present pickle and begin by setting up some definitions of the labels I have been tossing about and will be forced to continue using. Contrary to the older notion of a *humanist* as being one interested in the antiquities of Greece and Rome, the term here will mean one devoted to humane studies such as literature, art, music, and philosophy in the academic background, and one concerned with people and human nature in the case of the informed and curious layman. A *social scientist* deals with such subjects as sociology, psychology, and economics, plus other "ologies" on the university scene and off, while in the lay milieu he may work with non-academic types on such problems as transportation and urban affairs. I view the *scientist* as a man devoted to the pursuit of new truths about Nature and the universe, an *engineer* as one who, with due concern for time and economy, puts these truths to work, and a *technologist* as one who furnishes the tools and the know-how. I realize that there is, of course, tremendous overlap in the case of the last three, that the definition of the scientist might often fit that of a philosopher, for example. This leaves the poor *layman,* who seems all too often today to be at the mercy of all the others, but who, if he only knew (and many do), is right where the action is, has the greatest concern about modern science and technology, and, in the long run, can or should be a tremendous force in any program of sensible, profitable, or meaningful living. In the present context, for *layman* read *citizen*. All the rest are citizens too, of course, when they are not on the job—and thereby, too, as they used to say, hangs a tale.

The reaction to Snow in the last ten years has been a microcosm of the misunderstandings which because of faulty communication threaten the larger world or macrocosm in which we all live. Looking back now we may find, as did Martin Green, that Snow dismissed literary culture in England too cavalierly, Leavis wrongly dismissed social thinking as outside culture, and British literary figures opposed to Snow were guilty of "vested interest" anger. [22] But all reactions of this type were and are real. They go with the two-culture tradition. The extent of the semantic problem alone can be seen in the overlap of definitions of key terms given in the preceding paragraph. The more terms, the more apparent splits and splinterings; they are really only the exposed small portion of the iceberg, but large enough to sink the ship. (Ironically and paradoxically, the more splinters, the easier to recognize matching points as in assembling a jigsaw puzzle, and the easier to recognize that all are parts of one, except that few have had the patience to stick with the game.) Semantic barriers, clashing personal temperaments, different academic and professional training, propaganda and sabotage, hangovers from nineteenth-century tradition, succumbing to stereotypes, accepting isolated reports as typical, fear of loss of territory as in politics, unilateral instead of bifocal vision, over-intellectualizing and encapsulating, loss of basic feeling about people—these are some of the reasons for the gulf we have heard so much about. The sad thing is that both parties, scientist and humanist, are bright, decent human beings who find, if asked to take part in a panel discussion or if trapped at a cocktail party, that they have much in common, enough to start understanding, as in formal arbitration proceedings; with the latter, however, people are forced to sit down together for long periods of time—with the two cultures it is easier to drift away and melt into the security of a creature group with familiar plumage.

Part of the difficulty comes from hazy tradition, which dies hard, and part from ill-digested scraps of learning which circulate for the real thing. One man will remember Plato's remark in the *Gorgias* about the engineer being called an engine-maker, one whose son you would not want your daughter to marry. His opponent will remind him of the glories of the Renaissance when art and technology cooperated with God to finish building the universe. John Donne, seventeenth-century poet and priest, thought that new science put all in doubt, and the Romantics of the last century viewed the Machine, in the form of the locomotive, as a desecration of Nature's landscape, and the factory as a hell in which little children died from overwork and neglect. A tradition of aloofness sprang up which suggested that no gentleman need soil his hands on science, and many humanists today keep it alive. But literary reactions were also ambivalent, as we shall see in the next chapter: Whitman sang of the beauty and power of the Machine, Emerson found nothing inherently ugly about railroads, and even Thoreau found the sound of wind in the telegraph wires a new kind of music. Philosophers and religious folk blamed Newton for loss of purpose and Darwin for the exile of God, while today the man in the street reads an article or two and decides that automation is the devil's work. Auden reminds us that the "unknown citizen" had everything a man might want—a phonograph, a car, a radio, and a frigidaire—to which today we might add a TV, a second car, and a boat. He and his fellow poets wonder whether modern man is happy or whether he has not been dehumanized in a materialistic world made possible by technology. Yet even they would not want to give up wonder drugs or turn away from the abundance of food made possible by the same technology, and for humane reasons, as their opponents have been quick to tell them.

Time does strange things even to opinions like these, and a knowledge of history is again necessary, as Santayana pointed out, or we are doomed to go on repeating the errors of the past. (Item one on any get-together program: Let the artists and humanists read one volume of history of science or technology, and let the scientists and engineers read one volume of history of art or literature.) Notions about mathematicians, for example, show how opinions differ and how they change from time to time. Edward Gibbon tells us that he gave up math after mastering the basic principles before his mind became hardened "by the habit of rigid demonstration so destructive of the finer feelings."[23] The mathematician G. H. Hardy was shocked to find out that his interests disqualified him as an intellectual. But in 1967 Edmund Leach scrambled the picture by saying the following in the Reith lectures:

> Mathematicians have always been eminently respectable, and so are those who deal with the hard lifeless theories about what constitutes the physical world–astronomers, the physicists, the theoretical chemists. But the more a scientist interests himself in matters which are of direct human relevance, the lower his social status. The real scum of the scientific world are the engineers and sociologists, and the psychologists. Indeed if the psychologist wants to rate as a scientist he must study rats, not human beings. In zoology the same rules apply. It is much more respectable to dissect muscle tissues than to observe the behavior of a living animal in its natural habitat.[24]

This statement is interesting for at least two reasons: Anyone debating the two cultures would have different argument fodder depending on whether he quoted Gibbon, Hardy, or Leach; and phrases like "the real scum . . . are the engineers and sociologists" are so loaded as to do a great deal of harm, whether they are true or not. Squabbling between the two cultures is bad enough; infighting within either one, as here in science, makes matters messy indeed. And a further irony exists with regard to mathematics: in the highest echelons of this field many experts consider math not to be a science but one of the humanities, and they will discuss the aesthetics of an equation with great elan! The problem we are studying is not so simple as the phrase "two cultures" would suggest. And yet I feel that we can get nowhere until this question of images and attitudes is outlined in detail with examples and illustrations taken from past and present and from both sides of the fence.

Even a cursory study of images and notions of other men's occupations turns up a goodly measure of ignorance and nonsense. My own ignorance was colossal until I found myself teaching freshman English to embryonic engineers at what was then called Carnegie Tech. during the Depression of the early thirties. I had slipped through high school with algebra and geometry—no physics or chemistry—because I was in the college preparatory course! I satisfied the science "requirement" at Dartmouth while majoring in English by taking two safe and easy subjects, biology and geology, both of which were fun. Still no physics or chemistry, no knowledge of even the history of science. If this sounds unusual, it was not in those days; and, even today, there is less science required of liberal arts majors at Harvard than was the case one hundred years ago. My notion of a scientist was about that of many laymen today, nourished on TV commercials with their "scientific" demonstrations of dental plate cleansers: the man in the white coat, working in a room full of bubbly tubes in the best Hollywood tradition—a man to be respected, held in awe, sometimes feared, but never really known or understood. And, thanks to *Frankenstein*, it was not hard to succumb to the stereotype of the mad scientist, who all too often and all too easily was allowed to represent the whole breed. My first idea of an engineer was of the man who ran a steam locomotive. In college the engineers were often looked at as "different" somehow, not actually second-class citizens, perhaps, but scrubby mechanics with dirty fingernails who seemed to spend most of their time surveying a campus as if they were mapping the unexplored Northwest while they knew for the nth time where every marker was located. (In those days engineers got through with English composition or report writing and maybe some economics as their humanities background, so it should be easy for the uninitiated reader to see how class distinctions were born and why false images prevail.) If the engineer of my college days was not the man with line and transit, then he existed in an alternative embodiment, that of the man from the steam plant who came around periodically to tap on the pipes. Such is the manner of genesis of myth and superstition. As an English major I made one other encounter with engineering, in *Hamlet.* The line about the "enginer (sic) hoist by his own petard" could be relied upon to provide a moment's enjoyment in the long winter evening's reading; it was not until years later, however, that the full meaning of the unfortunate man's demise through the agency of his own land mine percolated through my mind in true perspective—after I had read the history of technology and learned about the glory of the ancient military engineer, almost a redundant term in the late Middle Ages.

As a result of teaching in three schools of engineering and science and one large university which had a flourishing engineering school and, most recently, after doing consultant work at another large university, my education progressed to the point where my prejudices became minimal and something like understanding prevailed. But most people who want a return to a unitary culture and who bemoan stereotypes and ignorance cannot invest this much time in reaching a state of grace. And even helpful education sometimes carries unexpected dividends, some humorous, some not quite happy. If one takes a postgraduate course in relations with engineers, for example, one runs into surprising roadblocks—such as their attitudes toward scientists and arts majors and vice versa, and the attitude of the casual public *toward* them. The American Society for Engineering Education in its April 1956 report on "General Education in Engineering" offers good evidence of the second attitude above:

> In specifying what is meant by 'misunderstanding and inefficiency,' we are compelled to report that the ancient war between the engineering and the arts faculties still con-

> tinues in a good many institutions, particularly the larger state universities, where the two faculties are frequently separated physically as well as spiritually. The arts faculty at such institutions still cherishes the belief that the engineer is a rough, uncouth fellow wearing boots and an open flannel shirt. . . . He has no manners, and he wants none. . . . His acquaintance with the arts is limited to cheap movies, and . . . with literature to comic books. He is crass, materialistic, insensitive. And it must be admitted that this picture appears to have considerable appeal for a certain type of engineer, who does nothing, in consequence, to destroy it.
>
> At the same institutions, the engineering faculty will give evidence that it regards the typical arts man as a pale, ascetic dreamer, forever in need of a haircut. He is devoted to modern art . . . music . . . literature, and he talks incomprehensibly about all three. He is addicted to books, and knows nothing whatever about science or mathematics. He is a thoroughly impractical fellow, probably a little pinkish politically, who gets by only because he has the gift of gab. And it must be admitted that this picture appears to have considerable appeal for a certain type of arts man, who does nothing, in consequence, to destroy it.[25]

There is something wryly humorous and something sad about these portraits. Again, the crowning irony is that if you place an engineer and an arts man on a design problem together they meet as friendly conspirators, sparks fly, and synergy results. Recent team experiments in Los Angeles involving teams of engineers and sculptors in metal proved to be exciting for all concerned—but ordinarily the taboos are hard to defeat. Having taught engineers, science students, and liberal arts majors, I recognize differences of personality, temper, purpose, and so on between groups and within them. But they all have good to high I.Q.'s; they all belong to the campus species of the human race, and they have much to give each other—if they don't inherit a batch of clichés and a tradition of prejudices the moment they enter the college gate. Labels get tossed around so carelessly. Perhaps the classic example of this with regard to engineers is found in the late H.L. Mencken's *American Language* where he discusses the American habit of using genteelisms (limb for leg), or euphemisms, or in other ways avoiding calling a spade a spade. A janitor becomes a custodian, for example, to dignify the job. In the case of the engineer, Mencken, leaning heavily on back files of *Engineering News-Record*, offers dozens of examples of misuse of the term, most of them humorous to anyone except an engineer and most of them, in their attempt to elevate, actually demeaning the original. Garbage men become sanitary engineers (this is familiar to most of us), but when workers in the shipping department become packing engineers, psychoanalysts become psychological engineers, men who attend to frisky bulls become dehorning engineers, and corn doctors become podiatric engineers, the labels cease to have meaning. I would agree with Mencken that the prize should go to "hot-dog-engineer."[26]

All of this provides fun for the arts people or the humanists, of course, and plays right into the hands of those who talk about a gulf between the cultures. It also provides fair game for those science types who like to look down on engineers except when they want to gang up to attack English majors. When Cape Kennedy was Cape Canaveral in the early days of experimenting with rockets and space vehicles, if a launching were successful it was inevitably hailed as a scientific achievement, but if it fizzled it was always an engineering failure, or so it seemed. Herbert Hoover, in his memoirs, tells an anecdote about a sea journey back to the States, during which he met a companionable lady with whom he got along very well conversationally. As the liner was about to dock, the lady thanked Hoover for making the journey pleasant, and then remarked that she had never asked what his occupation was. On being told that he was an engineer, she made the revealing remark, "Why, I thought you were a gentleman."[27]

Engineers, on the other hand, may feel that the scientists are glamour boys who make more money and have higher social status. They don't yet seem to realize that they are in a unique position to adapt science to human needs, and that they are moving up to decision-making of the highest importance. As for the scientists, studies of their image among college students reveal a mixed picture, again illustrating the power of stereotypes in producing friction among groups of people who really should be quite close. Beardslee O'Dowd, discussing this image of the scientist, summarized as follows:

> In summary, there emerges a picture of the scientist as a highly intelligent individual devoted to his studies and research at the expense of interest in art, friends, and even family. . . . He serves mankind in a selfless way, almost unaware that he is doing so; he serves others by serving himself. . . . He is clearly an intellectual, but unlike 'eggheads' in the humanities, he is characterized by a vigorous and directed use of his intelligence. The image conveys a sense of strength of personality, but it is a little extreme, a little strange, somewhat contradictory, and, therefore, hard to comprehend.[28]

How would a science student react to this description? How would a philosophy student react to "egghead?" Would the layman agree with the above?

There has been a good bit of arrogance on both sides of the culture gap, all the more disappointing because it comes from presumably gentle people with good educations. Casual use of language and epithet, as we have seen, intentionally or not, has a cleavage effect. A.G. Oettinger of Harvard, professor of linguistics and applied mathematics, has explained his purpose in writing on the jacket of his recent book, *Run, Computer, Run*: "I wrote . . . not as a Luddite fearful of the Machine nor as a shrinking humanist living in the past, but as a scientist and engineer convinced that educational technology holds great promise."[29] Is the implication that humanists shrink or that they live in the past a safe, let alone well-mannered, generalization? What kind of impression would such a gratuitous remark make upon a young science major?

On the other side, the humanist and the layman are guilty of contributing to what Professor Holton has called the "atomization of loyalties within the intelligentsia" and a loss of cohesion which is "perhaps the most relevant symptom of the disease of our culture." The ultimate cause? "A failure of image."[30] Among the distorted images described by Holton are those of the scientist as iconoclast, as destroyer of man and Nature and values, and as wizard whose magic often gets out of his control. He concludes that such prevalent images are the main reason for the split between elements of our culture and should be the concern of all of us. What is more, he feels that the split is beyond repair. I feel that he is definitely right about the false images, and can only hope that he is wrong about the repairs. At any rate, it is a great pity that in an age of super communication media, the communication itself should be so poor. What with arrogance here and inferiority complexes there and very little mutual understanding most of the time (as at the United Nations), one can appreciate Holton's pessimism. Not long ago, to get back to the engineers for a moment, I noticed this title for a paper delivered at a meeting of the Society of Automotive Engineers in Los Angeles: "Some Suggestions on the Management of Engineering Activity Based Upon the Radical Assumption that Engineers are Really People."[31] Consider the implications and tone, half-serious, half-joking, of that title and realize again the number of superiority and inferiority complexes that must be floating about, as documented in the last two paragraphs, in the fragmented cultures and sub-cultures of the modern intellectual-practical world.

The pity, of course, is in the fragmentation. And, at long last, some educators are getting around to this point. On December 15, 1968, the *New York Times* ran a story on Dr. Michel Crozier, head of a Paris research institute and former visiting professor in this country. Dr. Crozier, appealing for the applications of technology "within the context of a humanist perspective," said that it was "a kind of folly" to assume that "a rational view of the world based on the inevitability of scientific progress can cope with a fragmented, culturally diverse society full of complex emotional problems." He added that, by isolating itself from psychological and social insights, the technological approach is creating resentments and frustrations that, politically, have given rise to the beginnings of an anti-intellectual backlash among students and workers demanding a greater role for the individual in society. The next day, in the same newspaper, there appeared an interview with Charles B. De Carlo, the new president of Sarah Lawrence College. He too was quoted as saying, "I think we've gone too far in specialization, fragmentation, and technology. . . . If a college has to have a motto today, it ought to be 'Feel!'" His writings, the paper continued, argue for resistance of the dehumanizing aspects of technology by emphasis on the conservative, humane values of the liberal arts. All this is very encouraging, this deploring of fragmentation. But while Dr. Crozier ticks off the flaws in the rational view of a world based on scientific progress and President De Carlo talks of resistance to dehumanizing aspects of technology, both with a wish for less fragmentation, one can't help wondering whether they are not blind to the fact that their choice of phrase is not conducive to togetherness. Actually, what is needed is more tolerance and diplomacy on both sides, more sporting sense, better manners, really, but this is easy to say and hard to set up.

For now, I should like to give the last word on the interplay of images and attitudes that we have been discussing to J. Tuzo Wilson, quoted earlier. This seems to me to be a sane statement:

> We also know the humanist with the medieval outlook who pretends to have no patience with science, although often enough he is content to take advantage of its benefits. However, the technocrat who places all his faith in science is equally deluded . . . . In one lifetime a man can learn about only one thing really well. Each branch of human knowledge should be tolerant of the others and accept the notion that all parts are equally important. Just as the classical humanist loved the arts even if he could not himself paint or sing, so the scientist must learn to cherish his dependence on the humanities and the arts. The businessman and the lawmaker, too, must realize that scientists are not a group of outside experts to be called in for technical advice

> but are an integral part of present day society. Science has created a new philosophy of our place in the universe. It offers relief from the horrors of natural disaster and it promises an abundance that could bring plenty to all men.[32]

Enough of the thoughts of academics for the moment—what about the layman, the general public's attitude toward the exponents and the doctrines of the two cultures? We know that people in the United States have a mixed view of the professor as a type; in addition to some respect from the formally educated, there persist in other quarters, however, the images of the egghead, the eccentric, the absent-minded genius, or lately, the radical "nut." (Even the Winston cigarette advertisement poked fun at the caricature of a professor in mortarboard and gown who insisted on "as" for "like" in the famous slogan, "Winstons taste good like a cigarette should," you will recall.) The general public may not take to modern art or care for any modern poetry above the newspaper level, but they can take it or leave it. With particular regard for the much publicized modern scientist and technologist, however, there is no escape. The public attitude may be expected to be ambivalent, as was that of many nineteenth-century artists and writers, namely, one of mixed awe and far, or as Wilson again succinctly summarizes:

> The public may regard scientists as eccentrics, but it also tends to treat them with respect. . . . They are more often blamed as the inventors of the atomic bomb or the creators of unemployment through automation, than praised as the providers of modern medicine, abundant food, and the amenities of civilized existence.[33]

Insofar as the layman fears science and technology, he automatically lines up with many artists and writers on one side of the battle between the two cultures and, willy-nilly, knowingly or not, assumes a role in our present discussion; indeed, through his political power and his relation to the educational system, his may be the ultimate decisive voice. As the poet may show his concern over the exploitation and pollution of nature resulting from modern technology plus human callousness, the dramatist, the fear of the Machine and the Bomb, the philosopher, the threat of dehumanization in an assembly-line world, and the social scientist, the dangers of ignoring human values in a rush of half-thought-out urban planning, the informed and involved layman, though not as professionally articulate in some instances, has been vocal in protest, and what he cannot say, he reads, as the popularity of books by Vance Packard, Rachel Carson, Ralph Lapp, and Barry Commoner attests. Lord Snow may have had the academic and intellectual spheres in mind when he started the two-cultures idea on its way into circulation, but in this age of paperbacks any talk of man and science will now inevitably include the layman as well.

Things that bother the layman include waste, as in the planned obsolescence discussed by Vance Packard in the *Waste Makers* and other books; pollution, as described by Rachel Carson in *Silent Spring;* the breakdown of the myth of scientific or technological efficiency, as seen in the various efforts of Ralph Nader; and the disasters, threats of doom, poverty, and ghetto uprisings that furnish the daily headlines while astronauts land on the moon and the guns go off in Vietnam around the clock in an expensive nightmare of human ingenuity, know-how, blindness here, and misdirection there.

As Wilson suggests, the layman may be grateful for the things science can do (he certainly is glad for penicillin, TV dinners, and that new Buick), but he is also afraid of some things science, through engineering and technology, has done and may do. He is aware that air, soil, water, and even food have been polluted, and he is beginning to be aware of the problem of noise pollution. He has to go no farther than the *Saturday Review* to read articles with titles like "The Coming Struggle to Breathe," or *Fortune* with features like the late great mathematician Von Neumann's "Can We Survive Technology?"[34] Some of his friends remember the Tijuana kids who died from eating bread made from flour stored next to leaking containers of parathion or the Utah sheep which sickened and died near an Army gas depot. The loss of light and power in New York and parts of New England in November 1965 made him realize how everything taken for granted in the Technological Age cannot be taken for granted, that when the Machine stops the next time, as in Forster's story, we may all stop.

Ralph Lapp, in discussing "The Tyranny of Technology" in his 1965 book, *The New Priesthood,* comments on the power of modern science and the need of controls to save a society powerless to resist its charms. Controls would call for discussion of policy, but talk might have to be limited to those who knew the vocabulary. Here Lapp touches on a problem relevant to our present discussion of the layman's position:

> Just how far the dialogue can be extended to the populace is a moot point. The present educational attainments of the electorate make it difficult to establish good communication in scientific and technical fields. In the future, as the educational level is raised, the dialogue may be easier, but for the present it will probably have to be restricted to those who can understand the new jargon of scientists and technologists and to those who catch the sense of the conversation even though they are not literate in technical matters. After all, on the issue of water conservation one does not have to be an expert hydrologist to appreciate that the nation badly needs more water. Here the common sense of the average citizen still stands him in good stead. But when he is confronted with value judgments involving a choice of technological undertakings or a determination of the scale of competing projects, his common sense may no longer be valuable. Yet choices will have to be made—some on a monumental scale.[35]

The duty of the citizen is clear: use common sense, read to minimize ignorance, and work on that vocabulary. The alternative, as suggested by Bronowski, quoted in Lapp, ought to be frightening enough to stir even the most apathetic into action: "A world run by specialists for the ignorant is, and will be, a slave world."[36] Barry Commoner, in developing a similar line in *Science and Survival,* points out the involvement and responsibility of every man in an era which sees technology providing miracles at the cost of waste, pollution, and danger to health. Commoner thinks that those in charge have built a "magnificent material base of modern society" but also "threaten our very survival," largely through lack of care in thinking through the after and side effects of experiments and innovations.[37] Both Lapp and Commoner are more than scarehead journalists, and the layman should find no trouble with their language or with that of many similar books in print.

The layman and the humanist, the scientist and the engineer, the academic and the plain citizen, can and must join in facing the moral aspect of the issues stemming from Snow's far-reaching publicizing of the two-culture gap. To cite an article that anyone past grade school could read with appreciation, I mention Means' "Why Worry About Nature?" which appeared in the *Saturday Review* in December 1967. In discussing the effects of radioactive wastes, the killing of the passenger pigeon, the threatened extinction of seals and whales, plans for damming the Grand Canyon, and the pollution killing the fish life of Lake Erie, Means states the issue quite clearly: "Justification of a technical arrogance toward nature on the basis of dividends and profits is not just bad economics—it is basically an immoral act." René Dubos, in the same issue, urges more informed participation by all so that we may change "Where is technology taking us?" to "How can technology help us get where we want to go?" He simultaneously deplores the new image of the scientist and social complacency as seen in the notion that science can solve all problems by inventing new technologies and counter technologies; that one needn't really worry about cigarettes because science will cure lung cancer; and that overpopulation isn't really a problem because THEY will invent new foods and settle people on the moon or the ocean floor. Writes Dubos:

> Until recently it was believed that when technology created a new social problem a counter technology could be developed as a solution. But it is now becoming evident that in

> general new techniques with their counter techniques correspond to narrow-range, short-sighted adaptive mechanisms that will not long be able to take care of the problems of modern societies. Automobiles and super highways constitute caricatures illustrating the social failure of technology.[38]

Thus a biologist views us in our present state of evolution and adds a note of social responsibility to Means' moral injunctions. Finally, drawing together the topics of control, survival, rape of Nature, and social complacency discussed by Lapp, Commoner, Means, and Dubos, respectively, Harold Green looks at the new technological era from the viewpoint of the law. He sees our society committed to technological progress and is worried about the threat of destruction to human beings in the process. How much damage, he asks, could one demented or evil person inflict on society in one action twenty-five years ago? Today he might be responsible for snuffing out millions of lives. He raises the question now becoming increasingly prominent: Can our legal system impose effective social control over new technologies before they inflict substantial injury upon society? Green goes into radiation, sonic booms, weather modification, and bio-genetical engineering, all of which raise problems of manipulating our environment, and suggest that the basic need is for the long view (I would add wide) from the moment a new technological development commences. Pending legislation introduced by Congressman Daddario of Connecticut to establish a technology assessment board is a step in the right direction.[39]

In the past few pages I have reviewed and summarized the main points of Snow's work on the two cultures and some of the critical reaction. I have shown the persistence of the phrase in the language today and have gone into some of the problems Snow missed to bring him up to date as far as changes in education on both sides of the Atlantic are concerned. I have tried to indicate that the dichotomy which he publicized has vast present meaning in the direction choices our society must make. The problems of definition, of semantics, of image, and attitude were touched upon to show the difficulties of communication in the continuing debate. In a sense, this has all been summary and, in another, only preliminary. Much of it is academic. It is high time then to move on to something more explicit, more dramatic, a typical up-to-date confrontation: the record of what art and literature have had to say about the Bomb and the Machine, products of that "other" culture, since Hiroshima. Then, if Chapter I suggests the need of education as a solution to this continuing sparring match, we will study a typical experiment in curriculum planning. Finally, with the tradition reviewed, the specific confrontation, and a type of educational "solution" behind us, let us turn to arguments for the longer and the wider view, the demand for it, the reasons for it, the ways of achieving it, and the hopes for it, before both sides kill each other off and life becomes empty, mechanical, and terrifying—more so than many seem to feel it is already, judging by the contemporary protests, revolts, and demonstrations. But first, as they say on that technological contrivance, TV, but first, how do those pros, those poets, painters, novelists, and playwrights, those exponents of humanist culture, react to the Age of the Machine?

Chapter II

A Typical Contemporary Confrontation: Technology, Literature, and Art Since Hiroshima

"There seems to be no place where the cultures meet."
–C.P. Snow

1. Introduction

I may begin a discussion of post-war literature and art with the apologetic cliché that we are probably too near the woods to see the trees. And yet there is a path through the woods, and paths, like open doors, invite. It is not long before the signs and blazes look familiar. The statements of modern literature and art sound much the same with regard to the Machine as did those of a hundred years ago; the tone, however, is a bit shriller, more fearful, almost despairing at times. Indeed, these statements (distilled from a mash of style and media) sound familiar because essentially they are the statements to be found in textbooks, newspapers, articles, and other sources not ordinarily thought of as really literary or "artistic." The point is, however, that they are worth listening to; they deserve consideration if only in rebuttal, and, since presumably they represent the voices of the masses insofar as writers and artists have always spoken for and to the inarticulate, they deserve a measure of respect. But I am getting ahead of myself.

A sampling of literary and artistic materials to be discussed later quickly reveals, if you will pardon some overlap and disorganization attributable to the messy and disorganized mind of this itinerant humanist, the following opinions on Technology, the Bomb, and the Machine since Hiroshima:

Category I: Positive or Friendly Statements

1. M. offers new tools, materials, and images for sculpture.
2. M. offers new subjects, metaphors, and language for poetry.
3. M. liberates the human spirit by removing drudgery.
4. M. is a cultural symbol, a token of meaning and value.
5. M. today is a beautiful object, has its own aesthetics and sensuousness.
6. M. is a tool for God's purposes.
7. T. is sublime, power fused with Nature.
8. M. is a symbol of order and efficiency.
9. M. produces abundance.
10. T. is inspiring to writers of certain political groups, e.g., East German.
11. T. has produced medical miracles, may inspire new epics.
12. M. has positive virtues – law, order, discipline, energy – producing a city conglomerate for artists and writers to live and work in.
13. We can learn from the M. (e.g., brain function from studying computer).
14. The B. produced a moral reawakening.
15. It is nonsense to talk of scrapping the M.
16. The M. woke up the architect.

Category II: Equivocal, Ambivalent, or Questioning Statements

1. M. produces ambivalent reaction in writers: awe and fear.
2. Is M. to be instrument of human will or dominating god?
3. T. determines the direction of history and society.
4. M. should be a means and not an end.
5. Evolution of the M. will control evolution of people.
6. We must assimilate the M. and develop human possibilities.
7. Poetry is important to survival as a myth maker, since society is being hitched to the M.
8. Poets may work with technologists to open up a larger panorama of real problems of beauty and value.
9. Man must periodically escape M. and T. in Nature.
10. M. is not a monster if controlled; however, a technical coup d'état threatens to shape our lives.
11. M. is no longer an adequate word for a cybernetic world in which T. is less and less "mechanical."
12. M. has majesty even in its destructiveness.
13. Man needs to create a Whole, assimilating M. as a means.

14. The question should be, "Where can we lead T.?" – not, "Where is T. taking us?"
15. The danger lies in thinking that T. can do anything.
16. Can the M. ever be invented which will be concerned with truth?
17. T. enlarges choices, creates new dangers, blights, and benefits.

Category III: Essentially Negative or Unfriendly Statements

1. M. kills the personal self, individuality, unmans man.
2. T. has removed myth, magic, religion – all necessary to man.
3. Technological fact makes purpose, meaning, and the value illusory – hence an alien world with the M. as a symbol.
4. M. spoils the peace, beauty, and bliss of the natural landscape.
5. M. maintains a factory system, a modified serfdom.
6. M. brings material gain but spiritual, moral, and imaginative loss.
7. M. produces smog, noise, pollution: a health and beauty menace.
8. M. is a threat to arts and crafts because of mass production.
9. M. has no goals but its own ceaseless expansion.
10. M. rhythm destroys natural body rhythm and eventually mental life.
11. T. produces war machines and may end the world.
12. M. causes unemployment because of automation.
13. M. will replace thought, muscle, emotion.
14. T. leads us to conformity and an omnipotent political apparatus.
15. T. advances willy-nilly without guide plan and with results not needed.
16. M. has separated man from Nature and undermined his own nature.
17. The price of M. gadgets is servitude to a new priesthood.
18. Modern literature reflects the B. philosophy of death in life.
19. T. can alter human personality by experiment – human engineering a peril.
20. T. and M. condition people to forget the validity of non-scientific experience and non-verifiable insights, kill intuitive imagination.
21. T. has sold out to Big Business.
22. T. has swept man from his moorings; the B. produced existentialism.

23. T. has over-technologized writing and art.
24. T. has established a jerky tempo and wrecked our nerves, cutting us off from our biological heritage.
25. T. management can absorb our hostility to it by allowing for it.
26. T. produces boredom, loss of touch with work, loss of wonder.
27. T. emphasizes materialism, production, kills imagination.
28. T. norm is change and calls for a mass audience.
29. M. may soon make our decisions.
30. T. kills a fundamental human need: actual participation (methexis).
31. T. imperative is a dread of waste; a fulfilled life must have waste.
32. T.'s drive for accuracy is stifling.
33. Every T. development, even if not good for welfare, is in the totalitarian system to stay.
34. M. provides passive amusement, kills the creative, frustrates.
35. Modern T. communications wreck privacy.
36. T. makes us prize facts over ideas.
37. T. dehumanizes by evading surprise, hazard, grace, chance, unexpected choices.
38. The hold of T. is as alarming as was that of religious bigotry or serfdom.
39. TV encourages Americans to live a mindless life.

What follows in the body of the chapter is a condensation of reading in criticism, anthologies, primary works of fiction, drama, and poetry, and in articles of literary value on art; it will be followed, in turn, by what may seem a naïve, but what definitely is an honest, attempt to relate these preliminary findings to the two-culture question, to make hypotheses, and to raise questions based on a sincere belief that modern technology has something to gain from closer association with writers and artists (or their work) and vice versa.

A preliminary perspective on post-Hiroshima literature demands a frame of reference for modernism in art and literature in general. A useful summary of ideas can be found in "The Idea of the Modern," by Irving Howe, itself the Introduction to a book bearing the same title.[1] Howe notes that the Avant-Garde has risen as a special caste; the problem of belief has become more severe; "The idea of esthetic order is abandoned or radically modified"; Nature has virtually disappeared as a setting for literature; perversity has become dominant; the novel features a "whole new sense of character, structure, and role of the protagonist"; and nihilism "becomes the central preoccupation. . .at the heart of modern literature." One can find these notions in Dostoevsky, Flaubert (Life is so horrible that one can only bear it by avoiding it, and that can be done by living in the world of art), and Nietzsche, among others. But the modern hero outdoes his progenitors, believing in action but now unfit for it; keeping alive by running; living according to a private code; finding no meaning in life—in short becoming an anti-hero who salvages a portion of the heroic by facing this bleak nothingness. Howe concludes, "Nihilism lies at the center of all that we mean by modernist literature, both as subject and symptom, a demon overcome and a demon victorious. For the terror which haunts the modern mind is that of a meaningless and eternal death." Robert Brustein ("Who's Killing the Novel?") chimes in, "Among the reasons offered for this dearth are the Cold War, the Bomb, pornography, and the absence of style."[2]

According to Ihab Hassan, a Wesleyan professor-critic much in vogue, the writing, music, dance of today "betray a distrust of civilization and reflect the impulse to assimilate Chance into Pattern—all the characteristics of anti-form. . ." and, noting our "growing sense of collapse," he adds the following propositions:

1. Society, as a rational, historical, or moral order, is no longer available to the writer.
2. The new hero, a rebel-victim, projects a difficult and private existential ethic.
3. Form, by opening itself to chance and absurdity, invites harsh distortions, and sometimes verges on denying itself.[3]

And Charles I. Glicksberg sums up the prevailing general contemporary view as a "tragedy of modern consciousness. . . as man finds himself cut loose from his ancestral moorings."[4] He speaks for many who sense in the prevailing cosmic doubt, the feeling of aloneness, and the threat of cataclysmic violence, the evils of modern technology: ". . . so long as the atomic bomb throws its horrendous shadow over sublunary existence and the memory of Hiroshima anticipates what is to come for all of us, then existentialism is bound to take root as a philosophy of crisis." Not all contemporary writers see the same technological facts in this gloomy light, however; Elizabeth Sewell[5] reminds us that writers can supply the myth and the simple affirmation of the human body and mind, the only two methods of the imagination to face

> matters too enormous for our capacity. Such would include the enormities with which we have been confronted by science and technology within our lifetime. The first is Auschwitz. . . this first terminal point of our technological age. The second terminal point is Hiroshima. And in the literature about this (John Hersey's *Hiroshima*) we may see. . . the affirmation of simple humanity, the human being as sentient and conscious organism. . . .

Whether hopeful or despairing, poems about computers and plays about people surviving the hydrogen bomb seem to agree on the prominence of man's modern knowledge and intelligence. Charles E. Silberman ("Is Technology Taking Over?")[6] takes a sober middle course, but then he is one step removed from so-called creative literature, being a *Fortune* editor! He nevertheless offers modern art as defense, as new vision with the aid of technology, and sails into Ellul and those who prematurely mourn the dehumanization of man; Silberman counters with the notion that technology is neutral, that its impact depends largely on us, and then counter-counters with the admission that technology may not *determine* our destiny, but it surely *affects* it! In the long run, he points out, technology enlarges choice, also creating new dangers, but this is not new: long ago, Whitehead pointed out that the great ages by and large have been the dangerous and disturbed ones. Still, a note of skepticism prevails. And perhaps the difference between today's literary and artistic attitudes toward the Machine and those of yesterday is mainly a matter of degree, not of kind. The battle lines seem more clearly drawn—where yesterday any single writer might show ambivalence toward the new technology, today the cleavage for the most part is between writer and writer, artist and artist, with more emphasis on the negative reaction certainly. At any rate, many seem to share a feeling that technology may be moving ahead faster than man can keep up; that emotion, intuition, the creative spirit, and nonverifiable truths—the stock in trade of artist and writer—may be swept away; that "with man's invention of the H-bomb and his growing influence over his own environment, intelligence has become a factor threatening, rather than enhancing his chance of survival."[7]

So much for a brief introduction to a sampling of materials and expression to come. It is meant to set a mood for the reader, to furnish a little background music, if you will. And just as a professor of nineteenth-century literature feels that he must begin his course with a discussion of eighteenth-century writing, I feel that modern literary and artistic attitudes toward technology will make more sense if we go back, briefly, to studies of the attitudes of an age just previous to ours, then come up to the present, make some tentative conclusions and guesses, and end with some remarks on the possible relevance of all this to education and society.

2. General Background

A good place to begin the study of technology in literature is Leo Marx's *Machine in the Garden,* already a classic in its field.[8] Although Marx mentions modern writers, his main concern in this volume is with the intrusion of the Machine in the landscape of books by such nineteenth-century American figures as Whitman, Emerson, Thoreau, Hawthorne, and Melville. Here we meet the whole tradition of American pastoral, exploitation of Nature, emergence of the Machine as a cultural symbol, material gain versus loss of innocence, "the ancient war between the kingdom of love and the kingdom of power."[9] Most of the statements about the Machine made in my introduction can be documented in the early American classics as Marx interprets them—the mixture of awe and fear, aesthetic repulsion and commercial gain, so common in the press to-day. The author reaches the conclusion, after exhaustive evolutionary study, that through Machine domination "our inherited symbols of order and beauty have been divested of meaning. . . and in the end the American hero is either dead or totally alienated from society, alone and powerless. . . ."[10] Marx's thesis, only sketched out at the end, is that Frost, Hemingway, Faulkner "invoke an image of green landscape as a symbolic repository of meaning and value."[11] In a recent paper[12] he has suggested that urban planners give heed to what appears to be a basic human need for a spot of green somewhere in the background. And he writes to me that he is now at work on a sequel which will bring his studies up to the present. If and when this work appears, it should be a "must" item in any bibliography on the interplay between literature and technology; Marx is inclined to feel that while writers have shown and can show the situation and may be able to create new symbols, it is eventually a social and political problem, not one of art.

A recent work which should win a place next to *Machine in the Garden* is Herbert Sussman's *Victorians and the Machine,* 1968.[13] Sussman does the same thing for nineteenth- and early twentieth-century English authors as Marx has done for American. Among the figures under study are Morris, Kipling, Carlyle, Dickens, Ruskin, Butler, and Wells. Again, with new evidence, we find the Machine transforming the landscape, introducing social problems, stimulating aesthetic discussions and, as before, creating an ambivalence of opinion which, on the one hand, sings a joyous "Song o' Steam" (Kipling, "McAndrew's Hymn") and, on the other, produces Butler's view that Machines will breed and eventually people will be developed to fit the requirements of machinery.[14] Sussman's conclusion is as follows:

> In the machine, then, the Victorian literary imagination saw incarnated the power of the scientific intellect, and the finest writing on the machine centers upon the alliance of this abstracting intellect with the immense physical and social power of technology itself. The opposition to mechanized production, the celebration of hand labor, the aesthetic distaste for the machine are but expressions of the deeper conflict between rationalism and intuitionism, between scientific and organic modes of thought that is the true subject of the Victorian writing on the machine.[15]

This "deeper conflict," needless to say, is still going on. In reading about the Victorians in Sussman's brisk and yet scholarly treatment, one gets the pleasure of reading passages from well-known writers which are much to the point but, more important, also acquiring a sense of tradition and evolution so valuable in studying modern points of view.

It is difficult to unearth full-length works on modern or contemporary literature and its relation to technology, perhaps, again, because we are too close to the materials. Certainly there are few of the stature of Giedion's *Space, Time, and Architecture,* which treats of art and engineering, or Kepes' *New Landscape,* which deals with art and science. I can produce only two, and of these, only the second is post-Bomb, that is, as far as coverage is concerned; both are of recent date. The first is T.R. West's *Flesh of Steel: Literature and the Machine in American Culture,* 1967.[16] This is a flawed work, as they say, with an odd assortment of figures under scrutiny, at least two of whom are hardly "literary" people; however, it is worth a look. West uses "literature" casually as he lumps together Sherwood Anderson, Dos Passos, Sinclair Lewis, Veblen, Mumford, and Carl Sandburg, along with Waldo Frank, who is virtually unknown today to many readers. West's overall contribution is some coverage of American writers of the twentieth century, thus continuing, in a sense, where Leo Marx left off; he is, as stated, in no sense contemporary or post-Bomb. His main contribution is to offer a good argument for the fact that, although he concedes most of his (arbitrarily) chosen writers are anti-Machine most of the time, the Machine has positive virtues—law, order, discipline, and energy—which actually, perhaps at a price here and there, produce an urban conglomerate like New York, whose separate individuals are allowed to be artists, writers, and workers pretty much on their own terms, and who could not exist if the Machine stopped. He is closest to the Mumford of *Technics and Civilization* in suggesting that the Machine is assimilable and ultimately, with people also adapting, definitely not a monster if kept under control. Whereas Marx and Sussman discuss the ambivalence in attitudes of nineteenth-century writers toward the Machine, West prefers the term "polarity." Thus one can speak of the splendor of power and the aesthetic of the Machine versus its discipline, which establishes order but may repress individual experience. One comes away with the feeling that the modern writers are more anti-technology than their forebears, a feeling which grows much more pronounced after Hiroshima. As West sweats over the dilemma of modern environment ("Should that environment primarily be expressive of intuition, an artist's medium for the sensibilities. . . or should it be essentially a lawgiver [the machine, he has said, is the most logical of symbols for the fact of Law], resisting and toughening to some specific rule the anarchic materials of human temper?"),[17] he touches upon the general problem of the two cultures, always lurking in the background of the particular discussion of technology and literature.

The second of the two studies mentioned above is the very recent *Literature and Technology*, 1968,[18] by Wylie Sypher, chairman of the Department of English at Simmons College, whose critical work in both art and literature is known to and well thought of by both Marx and Mumford, for reasons which may become clear. Much of this book (after the opening section) is on art, as a matter of fact. Sypher points out similarities and differences between art and science, but looks upon technology with its system as an enemy of both. He turns to poetry and art as defense against the technological bogey man and essentially is asking for a return to crafts. He is quite erudite, but his style is often opaque, and he needs more specific examples, rather than generalities when he handles modern literature. The volume is helpful in continuing, as did West, the survey of interplay begun by Marx and continued by Sussman. Certain pages in Sypher are admirable presentations of a point of view. He makes a good case for the need in man to participate (methexis).

Among the points raised in *Literature and Technology* is the following:

> If alienation is a mark of technological society, it is also a symptom of the distancing that aesthetics presumed to be necessary to Art. . . .I have tried to suggest that the resolution of the technological difficulty may be a return to craft—which is not methodological.[19]

Curiously, Sypher's argument, which is modernized Ruskin plus a dash of Sir Herbert Read, is quite methodological. (His worst statement is, "I am assuming that engineering is hostile to art.") Sypher's main attack, however, is directed at the "technological imperative," which he defines thus:

> It is a dread of waste, a concern for efficiency that arises from a psychology of parsimony, or, to use the term we shall need later on, thrift. It is the principle of minimal effort, usually directed toward the solution of immediate problems instead of toward investigating the nature of the problems themselves.[20]

Against this the author poses the humanities and the "ethic of waste":

> . . . in the fulfilled life there is always waste—a waste of virtue, noble intention, intelligence, and effort. The technological axiom of equal, or commensurate return for expenditure denies the disproportion in tragic events, which cannot be made to balance. When man is able to realize his nature, there is an expense of spirit that cannot be justified by any equation he can write. . . .Art is a mode of waste, a desire to make something for the satisfaction of making it as well as for its utilitarian worth. . . .[21]

I will come back to this kind of discourse, quite characteristic of modern writers vis-à-vis technology. In all fairness to Sypher, he does admit toward the end that "Of late technology has been touched by the joy of finding in its solutions the play of intellect that satisfies man's need to invent."[22] His final word is a challenging glove slap in the face of the Enemy, which is not "technology but the official program," and his final hope is that "art remains our only refuge from a technological order where all can be calculated, formulated, regulated." He explains,

> In our technological culture the artist's vocation is resistance to human engineering, which is a perversion of technology. Sometimes his only mode of resistance is insolence. It is an insolence that can be justified only by considering that officials are even more colossally insolent in attempting to engineer human beings.[23]

Before turning away from general background to a consideration of technology as it appears in contemporary literature, genre by genre, I must call the reader's attention to one more work which might more logically lead off the next section but which more fittingly will end this one: *The Industrial Muse,* 1945, edited by J. Warburg. A collection of verse, it covers the Industrial Revolution from its beginnings and comes down to World War I and shortly thereafter; although one or two of the poets and poetasters included are still alive, the collection does not make the post-Hiroshima category. The volume may well serve as dessert after dinner, but it is no mere frothy meringue; social criticism and satire are mixed in with sentiment and praises of the Machine. All in all, it illustrates very well in another medium, which is indeed often the message, the sober analyses of the authors mentioned in the preceding pages. There are poems in praise of Savery's pump; poems objecting to ironworks violating the "muse-devoted vales"; lines praising steam, condemning factories ("the modern rack!"), weeping over child laborers, hymning the cable, the electric light, the threshing machine, the motor car, the airplane. Although the quality of the verse varies from good to unbelievably bad, it has real documentary value and is at the same time good fun, a welcome change of pace in the midst of trumpeting and debate. It is interesting, finally, to note the change in tone in this anthology as we approach modern times. Auden's "Unknown Citizen," with a number, not a name, has "everything necessary to the Modern Man, / A phonograph, a radio, a car, and a frigidaire." D. H. Lawrence ("Last Poems") sees natural life perverted in industrial society and the Machine driving man mad and blind. And Louis MacNeice in "Birmingham" almost brings us up to date with his description of factory chimneys "like black pipes of organs . . . on sullen sentry duty," waiting to "call, in the harsh morning, sleep-stupid faces through the daily gate." [24]

3. A Note on Technology and Poetry

I have chosen to approach contemporary literature from the standpoint of genres, or "types," rather than from that of history, or "survey," to use the jargon of college catalogues. Since the oldest genre is poetry, let us move into that area first. To begin with, a reading of Douglas Bush's Harvard lectures on "Science and English Poetry" is suggested;[25] though a bit dated, and in spite of the occasional blurring of "science" for "technology" which is common everywhere today, they make an excellent transition from the general works thus far considered to particular poems and poets. Another useful work, about the only thing of its kind, is Paul Ginestier's "Poet and the Machine,"[26] a commentary and anthology keyed exactly to the topic of this chapter. Ginestier tells us that

> One of the great factors of today's poetical creation results from the conflicts and problems of adjustment which have developed out of the superimposition of a metallic rhythm upon the psychic rhythm.[27]

His selections indicate that some poets accept the fact that "Bureaucratic organization is the quintessence of mechanization on the social plane,"[28] some adapt, and some run away. This editor does for French poetry what Marx, Sussman, and Warburg did for British and American poetry, although he is by no means confined to the French. Of one poem by Ben Maddow, called "The City," which shows a female clothing dummy "behind glass, untouched by human hand/ With plaster pubis, thigh and docile belly/

Lifting the admired fabric up for sale/ While the living long to wear her enameled eyes," Ginestier says,

> This is exactly the trouble. The machine has perverted our souls to such a degree that it has managed to reverse the archetype. Our modern Galateas who are really alive aspire to become statues.[29]

His thesis amounts to this: Modern reality outstrips poetic imagination; the poet is now a spectator verbalizing for other spectators. Sometimes the "mystery that our mechanical civilization seemed to destroy is rediscovered at the very center of the temple of life."[30] And sometimes, the reader gathers, the modern poet comes to grips with technology and bends it to his own use: James Kirkup, after witnessing a mitral valve operation in 1957, suggests the possibility of a new type of epic as he describes the skill of the surgeon, which is a thing of beauty:

> You with a curious nervous elegance laid bare
> The root of life, and put your finger on its beating heart.

The jury is still out on the question of poets versus technology, as we shall see.

The next step, in the interest of time, is to go to standard anthologies of post-Bomb verse, those which are most used in the universities. One such is Donald Allen's *The New American Poetry: 1945-1960.*[31] The poems in this avant-garde collection are about love, anti-Establishment, and various abstractions, use the wide, free line which may or may not show a rebellion against form due to disenchantment with the Machine Age, are occasionally dirty, muddy, or meaningless, but they contain no specific references to the Bomb or the Machine. Donald Hall's *Contemporary American Poetry,*[32] quite representative and one of the best of the widely-used texts, contains no mention of them either. It is simply impossible to trace here any general malaise one may encounter to the effects of modern technology. The poems deal with landscape, religion, friends, nature, love, memory, family, and other poets. As is typical of contemporary verse, few selections are reports on life in the older tradition; rather, they express feelings, occasionally obscure, and free association. Solemn, tight, they are a far cry from the earlier irony and private references of Eliot and his school. One could find more distress over science and its uses in Tennyson or Arnold.

One of the best anthologies today is *The Modern Poets,* edited by John Malcolm Brinnin and Bill Read.[33] Its preface states that "Among the works of modern poets, poems that give pleasure are far more common. . . than poems that present problems." (I suppose this would play down technology and such.) The topics covered by these poems cover wife, child, lover, mother, landscape, odd folk, baseball, mice, zoos, cycles, snails, stars, boats, hawks, pigs, leopards, ducks, dead birds, snakes, home, and church. One might say this is conscious retreat from the Bomb or defense of sanity before the Machine, but it would be difficult to prove. In the entire volume of over four hundred pages, the following only are noted for closeness to the present subject.

Richard Eberhart's "The Fury of Aerial Bombardment" (often reprinted) is more about God and man than the Machine. Daniel Hoffman's "The Seals in Penobscot Bay" is about seals passed by a destroyer on trial run: "They hadn't heard of the atom bomb/ so I shouted a warning to them." But this is almost playful itself, and the poem is actually a modern version of Ulysses and the Sirens. Isolation is a recurrent theme in the volume, and we read that modern man's alienation is partly due to the Machine world, but typical selections, like Stanley Kunitz's "The Science of the Night" ("My touch is on you, who are light years gone"), are actually timeless. Robert Lowell's well-known "For the Union Dead" does have one definite comment:

> There are no statues for the last war here:
> on Boylston Street, a commercial photograph
> showed Hiroshima boiling
> over a Mosler safe, 'The Rock of Ages,'
> that survived the blast.

Howard Moss' "The Gift to be Simple" is a tribute to Einstein, whose life contained the irony that his theories were related to the development of the atomic bomb and its relatives, though he hated misuse of science. The poem celebrates the man's gentleness, intelligence, and compassion as an "angel in ill-fitting sweaters"–

> But if he were remembered for the Bomb,
> As some may well remember him, such a tomb,
> For one who hated violence and ceremony
> Equally, would be a wasted irony.

Howard Nemerov, in "Boom!," does a poem on a news clip about church attendance booming because of a general material progress. He satirizes everything that is "full": churches, beaches, gas stations, and toilets! Today is not like the days of Job or Father Damien or when St. Francis worked for the birds. People pray for cars and boats and

> That it may never be with us as it hath been
> with Athens and Karnak and Nagasaki.

Already the Bomb site has become a muted historical metaphor. And so it goes, with poems on pigeons on statues, summertime, ferry boats, kids, even a bull. The poems cited above are the only ones related to technology in the modern sense, a small and uncharacteristic group. One senses no despair or doom as with one coterie of novelists, no tragedy not already universal, no deliberate absurdity (as with most foreign dramatists today).[34] Indeed, at this point the post-Hiroshima generation of poets seems to be heeding the warning of a fellow poet made at the time of World War I. Hart Crane had written, "For unless poetry can absorb the machine, i.e., *acclimatize* it as naturally and casually as trees, cattle, galleons, and all other human associations of the past, then poetry has failed of its full contemporary function."[35]

Chad Walsh, editor of another well-known text, *Today's Poets,*[36] is one of those who find it difficult to pinpoint specific influences:

> It is extremely difficult to generalize about the total impact of depression, war, cold war, hydrogen bomb, racial prob-

> lems, etc., on the poetry of recent decades. Shapiro's early verse was mostly 'war poetry,' but he never considered himself a 'war poet.' Almost every poet shows at least an oblique awareness of the shattering events of the century, but this awareness is more often expressed indirectly, through nuances of tone, rather than by odes on a soup kitchen or even elegies for Hiroshima.

In general, Walsh's anthology covers the same topics as the preceding volume. One exception is related to technology: Gil Orlovitz has a poem[37] called "The Rooster" which spoofs science and materialism—

> constipation
> will be solved by
> automation

Up to this point, the poets of today seem to negate Galbraith's recent remark in *The New Industrial State* that "We are becoming the servants in thought, as in action, of the machine we have created to serve," a pronouncement which echoes Emerson's "Things are in the saddle" and Thoreau's "We are the tools of our tools."

Then we run into a contradiction. M.L. Rosenthal of New York University, a recognized authority in the field, has written a book called *The New Poets: American and British Poets Since World War II.*[38] This volume carries on where his own anthology of similar title had left off; it is mainly critical analysis, but it contains generous quotation. Unlike Walsh, Brinnin, Hall et al., Rosenthal takes a definite stand on poetry versus the Bomb, the Machine, science, and technology. He feels that poetry since the war "has taken on a new coloration, in effect a new sense of unease and disorder. . . It is that feeling which Robert Lowell. . . calls 'our universal *Angst*'—a heart-heavy realization that remorseless brutality is a condition not only of the physical universe but also of man himself."[39] He speaks of "the deep, and literal, absorption of our age in the terrors of war":

> War is more than a theme or subject for modern writers. It is a condition of consciousness, a destructive fact that explodes within the literature as without it. . . simply the final step in the technological alienation of sensibility.[40]

It is no surprise, then, that Rosenthal scants the poets represented in the foregoing collections in order to document his thesis. He features his personal favorites of the confessional school, notably Lowell, and he quotes Williams, shocked at an injured dog in the street, "as at the explosion/ of a bomb, a bomb that has laid/ all the world waste." Speaking of a poem by Ted Hughes on the vicious habits of a now-dead pig, the editor states,

> A passage like the one just quoted would have been less likely to appear before the last war. Its bloodymindedness is a reflex of recent history, the experience of the Blitz, the Bomb, and Auschwitz—an expression of them, a recoiling from them, an approach to experience by way of their implications.[41]

Finally, Rosenthal sees the contemporary poet's situation as follows:

> Of course there are exceptions, but by fate and by choice the involvement of most of our vital poets is with the simultaneous exercise of an almost helpless identification and sympathy with the victimized psyche of the present cultural moment, and of the will to define the situation clearly, to insist on the poet's meaningful relation to it, and to seek the way to transcendent meaning in the process.[42]

Frank MacShane, on the other hand, discussing "The New Poetry" last fall in *The American Scholar,*[43] cuts right down the middle between our first-mentioned editors and Mr. Rosenthal. He writes,

> Technological changes at once blight and benefit our lives. With the new freedom comes the new barbarism. Faced with these conditions, the contemporary poet is both attracted and repelled by the world around him. Despairing, he yet continues to write. . . .

The point seems to be that the new groups can be formal or free, and write about almost anything from social conditions to California real estate. If Richard Wilbur writes a poem about a toad killed by a power mower or Barbara Howes about a dog killed by an automobile, which is the major concern—the accident of killing in life, or the Machine? In another generation the animals might have been killed by men or by horses. The lesson is that you can find just about what you want to find. When today's poets speak out on the Bomb or the Machine, they are, by profession or art, more articulate than most people, and anyone interested in the implications of technology on modern life could do worse than heed them.

And now, what do the poets themselves have to say when we get them alone? Let's begin with John Wain, novelist, poet, editor, one of England's "Angry Young Men." Introducing his own anthology[44] in 1963, Wain felt that "the world has been different since 1945" but that people were too exhausted from war to change poetry very much; in England poetry reached its nadir in 1946-50, and influence passed to America, for "that war had ended with the fearful savagery of Hiroshima and Nagasaki" with exhaustion and boredom balanced by guilt and fear.[45] Later, Wain feels, poetry began to build again. His own poetry, illustrated by *Weep Before God,* published in 1961, is outspoken against the Machine, as this passage from Section VI attests:

> Next sing the machine, our glory and disgrace,
> Celebrate its possibilities, and tremble
> At the cold fury of its many revenges.
>
> It began with metal.
> Metal hates flesh,
> Hates everything that has a beating heart.
>
> . . . .
> Come, shape an elegy for victims of the machine![46]

And Oscar Williams' collected poems, as reviewed in 1965 by Robert Spector, were dominated by "images of personal and spiritual loneliness in the Machine Age. . .a chromium age of false emotions." Williams' work is described as a "romantic protest against the debilitating effects of a materialistic civilization."[47] Just as Esslin, whom we shall meet below, divides modern plays into those *about* the Bomb and the Machine and those *reflecting* their influence in more general treatments of alienation and similar topics, we should mention that there is a category of poetry specifically about the Bomb, as well as the Machine. Space allows but few samples. One is the collection edited by David Boulton called *Voices from the Crowd, Against the H-Bomb,*[48] a production of the Campaign for Nuclear Disarmament. In it, James Kirkup does a poem about revisiting Hiroshima, only to find the city synthetically restored: "Here atomic peace is geared to meet the tourist trade." As he moves about, however, he encounters the memorial at the bomb explosion centre with its museum showing a blouse "polka-dotted with atomic rain, indelible" and the "cotton summer pants the blasted boys crawled home in to bleed/ And slowly die."[49] Among major poets, as we move on, perhaps Edith Sitwell was most affected directly. C.M. Bowra, in his book *Poetry and Politics: 1900-1960,* tells us about it:

> At 8:15 a.m., 6 August 1945, the first atomic bomb was dropped on Hiroshima, and this so appalled and horrified Edith Sitwell that it changed the direction of her poetry. On the actual event she wrote three poems, 'Dirge for the New Sunrise,' 'The Shadow of Cain,' and 'The Canticle of the Rose.' For her the bomb was more horrifying than the war.[50]

In this connection it is interesting to note that the Russian poet, Voznesensky, who distrusted the West, wrote a poem on the death of Marilyn Monroe in which he saw a portent of "a universal Hiroshima" and made the "dying actress forecast 'suicides, suicides everywhere.' "[51] So Hiroshima in poetry serves as propaganda, elegy, metaphor for man's darkest deeds, and opening chord for an exercise in transcendence, depending on the time, the place, and the poet. There can be little doubt that, one way or another, it has left its mark. Six years after the event the poet Delmore Schwartz saw it this way:

> In the unpredictable and fearful future that awaits civilization, the poet must be prepared to be alienated and indestructible. He must dedicate himself to poetry, although no one seems likely to read what he writes; and he must be indestructible as a poet until he is destroyed as a human

> being. In the modern world, poetry is alienated; it will remain indestructible as long as the faith and love of each poet in his vocation survives.[52]

But, speaking of poets, Archibald MacLeish, for one, is afraid that modern life has dulled our senses, even to the meaning of Hiroshima. "We know what happened at Hiroshima," he writes, "We know all this. But do we feel our knowledge? Could we ever *think* about risking the possibility of a world-wide atomic war as a matter of face or official vanity if we did?"[53] MacLeish sees modern man as maneuvered by TV and managed to the point of acquiescence by material-mongers. "The real crisis in the life of our society is the crisis of the life of the imagination. Far more than we need an intercontinental missile. . .we need to come alive again. . . .[54] Here one of our prize-winning poets resumes the ancient poet's role of mourner and prophet. He speaks for many, as is expected of one who sees and feels differently from many others. This voice of conscience cannot simply be dismissed.

Not all poets, however, feel so strongly on the subject of modern technology and science. Some use a lighter touch and suggest compromise. May Swenson suggests in "The Poet as Anti-Specialist" that in a "compartmentalized, overorganized, scientific age, his art can show man how to stay human." As a poet of some repute, she feels her function is to warn. She seizes upon the astronaut as the most timely of subjects and has a little serious fun with him (first noting that "In space there is so little space"):

> Hooked to the indispensable members of his team by the paraphernalia of intercommunication, the astronaut, I imagine, must learn to forget what solitude, what privacy tastes like. . . .First trained to become a piece of equipment; next, perhaps, born so.[55]

And yet Miss Swenson notes a kinship between poet and technician which is partly vision and partly intent, and which also needs exploring in spite of semantic barriers. Both poetry and science, she says, try to make models of infinitude, science more demonstrably. Both use and rely on language; their impulses are parallel, although their instruments, methods, and effects differ. Though science is objective and poetry subjective, and science is commonly linked in the popular mind with reason, fact, and the material, while poetry suggests intuition, essence, and the spiritual, they are not as far apart as stereotyped thought would have it. Perhaps this kind of bridge-attempt is an answer. Another poet with a light touch is Peter Viereck, who represents the modern breed of professor who also does verse. Discussing the moral issue raised by the Machine and the Bomb in "The Poet in the Machine Age," Viereck classifies anti-Machine groups as including "esthetic wincers," the Lake Poets and the Pre-Raphaelites, who disliked physical ugliness; the "pious scorners" like Ruskin; "back-to-instinct prophets" like Lawrence; and "trapped individualists" like Arnold, who had deplored the danger of faith in machinery "as if it had a value in and for itself." In the pro-Machine category, Viereck includes "middle-class materialists"; capitalists, who link the Machine with God's work; "socialist materialists"; "gadget cultists"; and "lion tamers" like Emerson and Hart Crane, who use the Machine for ethical or aesthetic ends. Though directed for the most part toward pre-Hiroshima poets, Viereck's phrases serve well enough as labels for much of contemporary verse and versifiers, and his barbs thrown at all materialists getting fat on technology today are in the armory of poet-critics now writing. The war between love and power is timeless.[56]

To complete the gamut, one can find many commentators on poetry welcoming science and technology to the fold, for they offer new subject material (airplanes instead of skylarks), stretch the imagination, and redefine reality. Among them is Norman Holmes Pearson of Yale, who represents a sensible approach to the record of science, technology, and poetry in his article "The American Poet in Relation to Science," which appeared in *American Quarterly* after the Bomb. He is grateful to science for revitalizing poetic diction. He comments on the fact that science developed "a technology which became a comforting ritual for observance and endowed its own name with talismanic powers." However, Pearson is unhappy about the loss of individuality for both technologist and poet in their respective vocations in modern life. He brings the whole subject of this section into final summary and perspective about as well as it can be done:

> In a world which is unwilling to correlate the ways to knowledge of the scientist and poet, of reason and the imagination, the whole man has been lost by division. Personality is being erased as the area of reality is reduced. But if we look at the psychological history of twentieth-century America, it has been the poet who, acting on the concept of intuition, still penetrating the mystery of both the physical and metaphysical as they affect himself as man, has made the strongest public stand for the dignity and freedom of the individual. He has been the preserver of the Renaissance heritage. In his concept of total reality he has accepted what science has taught about this newest of new worlds, but in presenting it in a work of art he has encompassed an even wider sphere. It is at least possible that the American poet of the first half of the twentieth-century may be the complete realist, though not necessarily the complete mirror of the temper of our age.[57]

Fittingly and quite recently, the physicist who is also a poet, J. Bronowski, has again attempted "to correlate the ways," to restore the whole man, man as a unique creature, by pointing out the duty of science "as a humanistic discipline" to transmit this sense of uniqueness, to teach the world that man is guided by self-created values and thus comfort it for loss of absolute purpose.[58] And technology, as Sir Eric Ashby has indicated,[59] can be the cement between science and the humanities, in the van of which walk the poets still.

4. . . . and the Drama

Drama, unlike poetry, is slow historically to speak out on the subject of the Machine. In England, the eighteenth and nineteenth centuries are low spots indeed in the whole story of playwriting; in America, drama does not come of age until O'Neill. Ibsen and his follower Shaw were actually the first since Shakespeare to restore the theater as an arena for real discussion of serious issues. Shaw took an incidental poke at medicine and munitions in the course of writing the human comedy, and O'Neill in *Dynamo,* a lesser play, showed some awareness of the Machine, but only as one feature of a complex system run by the Interests. A good starting point for studying interplay between modern technology and drama would be the end of World War I, and three plays stand out in this connection: *Gas,* by the German, Georg Kaiser; *R.U.R.,* by the Czech, Karel Capek; and *The Adding Machine,* by the American, Elmer Rice.[60]

In *Gas*, as Ronald Peacock observes, "the theme is the dehumanizing influence of technocratic organization."[61] In the play an explosion takes place in a factory. The hero feels, like the public with regard to the New York power blackout or the Ohio River bridge disaster, that if Machines and structures can break down, he had better recover a human sense of values of pre-technological type. He feels that man is becoming enslaved in a society running by technocratic compulsion, that "Every person is chained to a function in a closely articulated mechanism; and when human creatures exist as no more than a function within a whole, the whole itself is not human."[62] The protagonist goes down before the demand for production and is unable to sell his gospel to the workers.

R.U.R. is a "comedy of truth" about the manufacture of robots to take over the drudgery of labor from human beings; they eventually improve by mechanical evolution to the point where they take over, killing their human masters. Domin, the factory manager, defends technological utopia because he wants man to become the master: "I wanted not a single soul to be broken by other people's machinery."[63] Capek thought him right. But, said Capek,

> Alquist. . .believes that technical progress demoralizes him, and I think he is right, too. Bussman thinks that industrialism alone is capable of supplying modern needs; he is right. Finally, the Robots themselves revolt against all these idealists, and, as it appears, they are right too.[64]

Rice's play deals with people who, for the most part, have numbers in place of names. The hero is Mr. Zero, an accountant who is fired when the Boss introduces the new adding machine. In a series of fantasy scenes, using the then-new technique of expressionism, Rice shows Zero electrocuted by the state, transported to the Elysian Fields, where he does not fit in because of his regimented ideas, and finally sent back to earth for another reincarnation. We discover that he has always had the mark of the slave on him; historically, in other embodiments he has been galley slave, serf, and pencil pusher. In the past he at least had muscle. Now he is a weakling, lost in the social assembly line. His latest task, and here the play ends, will be to operate a super-adding machine in a coal mine; once a day he starts the mechanism, which records daily output for each miner, by pressing a button with his big toe. He has no other task. The irony is that he thinks this is great. He has really arrived.

The tone of drama after Hiroshima, whether directly about the Bomb or the Machine or indirectly affected by the march of technology, is much sharper, shriller sometimes, disillusioned certainly, often defeatist, often violent for violence's sake out of sheer disgust. Martin Esslin of the B.B.C., author of *The Theatre of the Absurd,*[65] writes me,

> Of course there is a difference in dramatic literature since Hiroshima. One could subdivide this as follows: (a) general treatment of mechanisation, alienation in a machine-ridden society; (b) direct references to the atomic bomb and its effects. Under (a) there is a large number of plays about people being dehumanised. Ionesco's *Rhinoceros* is a case in point; admittedly there have been plays like that since Kaiser's *Gas*. . .but the current trend is far harsher. It is not easy to give really conclusive examples, as this is a *mood,* a general pre-occupation which pervades *all* serious drama.[66]

Under (b) Esslin lists some plays which I shall come back to and also notes the importance of contemporary Eastern European dramatic literature (notably in Poland and Czechoslovakia) "which is mainly concerned with bureaucracy as an aspect of the dehumanizing mechanization of man, and ultimately goes back to Kafka." Some critics will have none of this rationalization and find the dramatists of the Absurd a weak lot, as, for example, Glenn Loney:

> The unlucky Greeks had no such toys in their attic as the Hydrogen Bomb, but that did not prevent the great tragic playwrights from portraying man in circumstances quite as doom-laden and depressing as our own uncomfortable age. Unlike the writers of the Absurd, however, the Greeks did not throw up their hands in despair, babbling cascades of inverted clichés. Oppressed by unreasoning Fate, man still had dignity, and, after disaster, there was even hope. In my view, the theatre of the Absurd, in its defeatist celebration of emptiness and despair, is very much a function of our era, though it derives many of its techniques and much of its nihilism from the theatre and philosophy of earlier periods.[67]

Alfred Kazin would agree and add that

> Art has become too easy, a gratification of the self—and the self is terrible in its demands on our taste. Surely it is not just terror of the Bomb and resentment of racial inequality, militarization, and widespread poverty that leads so many commonplace and merely fashionable minds to echo the disgust with civilization sounded by Genet, Beckett, and other genuinely idiosyncratic writers . . . We are also propelled by the dissatisfaction created by too many easy satisfactions."[68]

Leonard Pronko, a Pomona professor and author of *Avant Garde: The Experimental Theater in France,* is interested in more general intellectual interplay and offers, also in a letter to me,[69] further insights:

> It seems to me that the post-bomb theatre is less apparently dealing with technology than the theatre before Hiroshima. One reason for this is, of course, that the terrible consequences of the bomb(s) posed some troubling questions, or perhaps put those eternal troubling questions into a new light. Consequently, this theatre tends to deal not so much with man in the technological world, but man within the metaphysical universe. For example, place Beckett next to Rice or Capek and you'll see what I mean. . . .
>
> What I've said applies, of course, to the serious dramatists of our day. There continue to be. . .spoofs and parodies of industrialized, mechanized society, but these plays

> (or films)...are the fluff of the entertainment industry most often....Therefore serious dramatists since the bomb tend to place the problem into a much broader (and vertical) context.

It should be pointed out that Esslin, in his major work, brought out that previous theater had reflected moral order from the mystery plays to Shaw–faith in Christ or evolution or progress. We do not have a common world picture today, he states. The playwright cannot proceed within a framework of values because the foundations are eroded by decline of religious faith, destruction of belief in automatic social progress, and discovery of irrational forces in the human psyche. Add to these a fourth reason: "the loss of a sense of control over rational human development in an age of totalitarianism and weapons of mass destruction."[70] It is Esslin's thesis that those playwrights who cannot do studies of Marxism, psychoanalysis, nature, or aestheticism in lieu of lost values have to turn to dreams, wild ideas, cruelty, the grotesque, non-sense–the Absurd. I would add, because drama is more direct, generally "easier" than poetry, both visible and listenable, it lends itself more satisfactorily to exposition of ideas and certainly, in the case of literature and the Machine, to study of tangible effects. The general disenchantment with modern life reflected in Esslin and Pronko's statements can be clearly seen in the following speech, for example, from Osborne's much-discussed *Look Back in Anger:*

> I suppose people of our generation aren't able to die for good causes any longer. We had all that done for us, in the thirties and forties, when we were still kids. There aren't any good, brave causes left. If the big bang does come, and we all get killed off, it won't be in aid of the old-fashioned, grand design. It'll just be for the Brave New-nothing-very-much-thank-you. About as pointless and inglorious as stepping in front of a bus.[71]

Significant dramatic reaction to the Bomb has come from England and the Continent, while American drama of any kind has hit rock bottom in recent years. Since early Miller and early Williams we have had nothing first-rate or consistent from anyone except Albee, and he has had three failures in a row. Off-Broadway and off-Off-Broadway have attracted attention for violence, pornography, and posturing—interesting for an evening, perhaps, or for drama historians, perhaps relevant in reflecting the malaise of a material world, bolstered by technology—but not well written and really with very little to say. Most Broadway plays fold after short runs; the survivors are mainly musicals, imports, translations, or fluffy domestic comedies headed for the movies. The reasons for the failure of serious theater in America are many—cost of production, local competition from repertory, union demands, Hollywood, and so on—but this is not the place for such analysis.

There is no lack of foreign plays on or influenced by the Bomb. In addition, there is a plentiful supply of drama criticism on both sides of the Atlantic, in book and periodical form. Glynne Wickham, for example, discusses Pinter's play-portraits of the individual in a society of mass conformity with the H-Bomb as its symbol: "In a world of science and related technology the right hand has effectively ceased to know what the left hand is doing—the world in fact of Harold Pinter's plays or Eugene Ionesco's."[72] Wickham refers to the tyranny of science and technology as being "as alarming in the hold which it has taken on men's minds, as those of religious bigotry and feudal serfdom which we have at last thrown off."[73] The author quotes Robert Bolt *(Man for All Seasons),* who compares the Deluge, an act of God, with an atomic war, an act of Man, who seems bent on self-destruction. The dramatist Bolt admits that the question of Man's nature has been a vexed one since he began to turn from God in the Renaissance, but "because the bomb enables Man for the first time to realize irreversibly whatever fantasies of evil he may have, we are the first generation which cannot dodge the question."[74] Robert Ardrey, the American dramatist, author of *Thunder Rock,* is similarly gloomy. He finds that modern plays, music, and art are all anti-story, anti-hero, anti-novel, anti-philosophy, to be found in non-books. The post-Bomb arts show "the neurotic intellectual attracted and absorbed by defeat."[75]

To the extent that plays reflect life, what can we expect, asks George Boas, if horror outside begets horror on stage?[76] We have had horror in Greek myth, in Christian martyrology, in Gothic novel, in crime fiction, but, Boas says, as he traces the evolution of the tragic hero, nothing like the "new crop of horror in the work of the Existentialists." This, of course, reflects the weapon which "may well exterminate the whole race." John Gassner, well-known theater historian and drama critic, writing on "The Possibilities and Perils of Modern Tragedy," echoes Boas and recalls MacLeish:

> How should the grandeur of the tragic hero and the splendor of tragic vision survive in a world levelled down by democracy and cheapened by mass-production and mass-consumption, a world in which even emotions and ideas have been converted into commodities gaudily packaged for the buyer?[77]

It is well-nigh impossible in modern literary and art criticism to get away from this charge that mass production, the child of technology, has killed taste, choked individual expression, and otherwise debased human life.

I have referred above to Harold Pinter, whose play *The Caretaker* has been made into a successful film, and who has, at this writing, two plays on Broadway. "The theatre of Harold Pinter," writes Bernard Dukore, "is one of the strangest types of theatre to have emerged during the atomic age." Dukore's description may stand for the world of John Osborne as well, for the stage world of Ionesco, Beckett, Genet, and Adamov:

> It is a picture of contemporary man beaten down by the social forces around him. It is a picture of man without identity and without individuality, of man crushed into a rigid social mold. It is a horrifying picture of contemporary life. It is a picture of the powerlessness of modern man, and the plays are frightening. It is a picture of the absurdity of the human condition in our world, and the plays are comic. But beneath the laughter and overpowering the laughter, there is a cry of despair from a well of human hopelessness.[78]

The theater of John Osborne and the other "Angry Young Men" started a new movement in Britain marked by plays of revolt, social criticism, and new post-war realism. Ossia Trilling says Osborne dropped a "dramatic H-bomb on the complacent middle-class theatergoer in April 1956." (Note how the Bomb has crept in as a figure of speech as well). Trilling's analysis of the new realism is of a compound of two elements:

> The first represents an involuntary and instinctive resistance to the irksome class structure of British society, and the other, in one form or another, the universal dilemma which has split the world down the middle in an age of threatening nuclear destruction and exacerbated the rivalries inherent in class, in nation, or in industrial or commercial power . . . at the heart of it all there lurks the unmistakable refusal of the common man to be put upon by the mumbo-jumbo with which the panjandrums and moguls of the new society are seeking anew to enslave his free spirit.[79]

In the interest of space-time, I shall mention one more critic, give two or three specific examples from the plays themselves, and hurry on to a consideration of fiction. Dramatic reaction to the Bomb and the Machine, in both plays and criticism, clearly documents the generalities laid down by Esslin above. George Wellwarth, in *The Theater of Protest and Paradox,* 1964, joins the chorus in generalizing from the particular case of the dramatist Arthur Adamov:

> Like most modern dramatists, Adamov has a thoroughly pessimistic outlook on life. His themes are the tyranny of parental love; the innate cruelty of the society in which we are forced to live; and the meaninglessness, futility, and confusion of everyday life.[80]

Wellwarth discusses Dürrenmatt, the popular and successful Swiss, who has two recent plays relevant to my subject: *Operation Wega*, 1954, and *The Physicists,* 1961. In the former, ministers from America and Europe go to Venus in space ships, to get the Venusians to side with them against Russia. Venus, strategically located, has been a garbage dump for political exiles. The mission is under orders to destroy the planet with cobalt bombs. In spite of constant storms, earthquakes, radioactive seas, and roaming monsters, men are free and too busy to make laws. The first play ends like the movie, *Dr. Strangelove,* in reluctant disaster accompanied by sardonic comments on the East-West struggle. The latter play deals with the dilemma of a physicist who has worked out a formula fór all possible discoveries. Horrified at what politicians might do with it, he has had himself shut up in a lunatic asylum with some real nuts, spies, and so on. The dialogue is rather good. The lesson of the play is that mankind can be saved only through suppression of technical knowledge. We have heard this before, hinted at in the public prints. But there is a difference between exposition in straight prose and in a play. Again, the medium is the message, a dramatic one, with a tone pretty much unknown before World War I, but with power-packed amplification (plus some distortion) since Hiroshima.

Perhaps the message of post-Bomb drama can best be conveyed directly by the plays themselves. Following are only three examples which must stand for the lot. Marghanita Laski's play, *The Offshore Island,* 1959, created quite a flap in England.[81] In it, Russia and the United States have been at war for ten years, using high-explosive, but not atomic, weapons on each other. They have, however, used them on the rest of the world for various political and tactical reasons. They have checked out spots where people have survived because of remoteness, caves, waterfalls, etc. They sterilize and transport survivors, known as C.P.'s, or Contaminated Persons, so that monsters will not be bred. The play concerns a British mother, son, and daughter who have survived, and describes their ingenuity at keeping alive. Americans and Russians appear during the course of the action and eventually agree to drop a small bomb on the spot. The son is killed when he resists the Americans, the mother, refusing to leave, awaits destruction, and the daughter flees. It is a moving picture of the real possibility of people fighting on and destroying without end or reason. Sample dialogue:

Charles (American officer): You're too intelligent for your own good. I'll tell you the truth. I don't know either. If *they* bombed you, it was because you were allied to us. If *we* bombed you, it was to forestall some move or other on their part. Little countries got bombed for all sorts of reasons, and which it was in your case, I honestly don't remember. After all, it was ten years ago.

Rachel: But we were friends.

Charles: Military considerations are more important than friends.[82]

In Robert Bolt's play *The Tiger and the Horse,* 1963, the Bomb is quite prominent.[83] The setting is an English college town; events concern a master's family. Students petition against the Bomb. Curiously, Mrs. Dean says, "These atomic explosions; have you noticed how beautiful they are? I didn't sign the petition against the Bomb, because I want the Bomb to happen." There are many arguments over the weapon, whether anyone will really use it, whether petitions accomplish anything, whether personal involvement as moral gesture is effective, and so on. Mr. Dean's defense for his wife's deranged act in slashing the college Holbein is pertinent:

> I am going to say that the purpose of the action was to publicise the petition, that my wife and I had allowed the subject of, er, nuclear warfare to prey upon our minds to a possible obsessive extent, and that we did it together![84]

The pervasive effect of the Bomb on writing, generalized above by Glicksberg with regard to poetry and Esslin on drama, is here made explicit.

Doris Lessing's play *Each His Own Wilderness,* first presented in 1958, is interesting as an early effort in the new post-war theater in England, dominated originally by Osborne and Simpson, then by Pinter et al. As a play, it is much too talky, with no Shavian wit to make up for lack of action. Still, the lines would interest a sociologist or a historian; the issues are still around. And the Bomb is there all the time. The mother, Myra, is an active anti-Bomb worker, going to meetings, handing out literature, taping sound effects. Posters ("Ban the Bomb!") lie about. Son

Tony is of the new disenchanted breed, like the speaker in Osborne's *Look Back in Anger.* Says he, "You're so delightfully old-fashioned. Getting killed for something you believe in is surely a bit of a luxury these days? Something your generation enjoyed. Now one just–gets killed."[85] There are various attitudes toward the Bomb: A politician says, "People who object to the hydrogen bomb are simply neurotic!" A churchman says, "The hydrogen bomb must be regarded by true Christians as part of God's plan for humanity." Myra wonders whether the blow-up of the human race would be such a loss; she has a recurring dream of the world as a large plain, filled with enormous black Machines. Milly says to her, "Ho-ho–so we're against the machine now are we? Back to the Golden Age!" It is Milly who puts the Bomb in proper cynical perspective:

> Why shouldn't we all ring each other up in the middle of the night and report progress. The grunts and groans of pleasurable love-making would be interrupted for the sake of a few minutes' militant conversation about the dangers of the hydrogen bomb. Then back to what everyone's really interested in. It's bloody funny, when you come to think about it. . . .[86]

Bloody funny, indeed. Behind the pun and "overpowering the laughter," as Dukore has said above, "there is a cry of despair from a well of human hopelessness."

I have not been able to find any plays that are happy about the Machine world or less than despairing about the Bomb. Naturally, many writers avoid the subject, and others do little comedies to take our minds off unpleasantness. Among serious plays, however, there is almost universal gloom or cynicism or cruelty or violence or nonsense, as we have observed, and to a degree unrecorded before Auschwitz and Nagasaki. Now and then a voice tries to strike back, but it is merely an objection, a bleat, followed by no program. The eminent critic and historian of the theater, author of the outstanding *Idea of the Theatre,* Francis Fergusson, tried hard in a recent article on the Theater of the Absurd, called "After Paranoia, What Next?" Fergusson is fed up, he says, with the "playwright-as-sadsack" and the "real lowdown on 'the chaos of the modern world.' "[87] So are a lot of us, maybe, on both counts. But to the question raised, "What next?" neither he nor we seem to be able to come up with a sufficient reply.

5. . . .and the Novel

If there were world enough and time I would be tempted to talk about technology and fiction, including the short story, but I have settled for the novel, which is a more comprehensive form, a form more widely read, probably, and certainly more advertised and reviewed. On some other occasion we might examine such story pioneers as Wells' "Lord of the Dynamos," 1894, which shows the Machine as symbol of both power and amorality; in it a scientist trains a savage to run a Machine, which appears to the latter as a god; in the end, however, simplicity is corrupted and the native worshipper destroyed. Again, no treatment of man and Machine could long ignore E.M. Forster's classic, "The Machine Stops," a model for later works by others; in it people live underground, with life maintained by the Machine (there is even a bible of sorts, the *Book of the Machine)* until one day the apparatus runs down, the people rush to the surface in time for one glimpse of blue sky, and die because they are not adapted to the atmosphere of the upper world, where a few creatures have apparently survived and will carry on. Another pre-Hiroshima tale with prophetic insight is Stephen Vincent Benet's "By the Waters of Babylon," in which nomads who have survived the great catastrophe come to what is left of New York and try to figure out what men or gods could have been commemorated in the cluttered stone blocks and fragments of Wall Street buildings, one of which spells part of "Washington." Among post-Hiroshima short fiction, I may cite Walter Van Tilburg Clark's story, "The Portable Phonograph," an account of a few survivors of the Bomb who huddle together and on rare occasions listen to an antique crank-up machine whose few intact records are cared for like diamonds or rubies. The tale raises an old question: although most of humanity has been killed by a Machine product and discussion among the survivors damns Machine civilization, they are dependent on a Machine to preserve their sanity. Again the intellectual Luddites face the dilemma: what Machines will you keep? Finally, in a more macabre setting, we could hold forth on horror stories of the kind represented by Kurt Vonnegut's recent *Playboy* specimen, "Fortitude," which tells of an old lady who, over the years, has had most of her organs replaced by mechanical transplants made by Westinghouse and others—"she had the best set of sweetbreads that money could buy."[88]

A researcher looking through *Reader's Guide* from 1945 to 1968 under "Novel" is immediately taken with two facts: many critics feel that the novel is sick, moribund, dead, or in transition no one knows where, and many disagree; a large and vocal minority of novelists feel that the world has gone to hell because of the Bomb, the Machine, political corruption, materialism, loss of personal identity, and other forces often traceable to technology, and the rest are going off in all directions, business as usual. In discussing the extreme view, Ihab Hassan ("The Novel of Outrage") reviews Styron, Baldwin, Ellison, Mailer, and Burroughs. In treating the controversial novel, *Naked Lunch* (1958), by Burroughs, Hassan says, "He offers a deposition against the human race, a testimony of outrage in the metallic voice of a subtracting machine." And Burroughs himself represents these best-selling writers by suggesting the redemption of creation by abolishing it; his only alternative is to "Reverse all your gimmicks .... Reverse and dismantle your machine."[89]

A bit of perspective would be helpful here, and Alfred Kazin comes to our aid in an article called "Imagination and the Age." He reminds us that in the 1940's the early modern novelists – Anderson, Dreiser, Cather—began to drop off. Saul Bellow's first work came in 1944, the year of *Glass Menagerie*. In the Bomb Year of 1945, *Cannery Row* and *Animal Farm* appeared. This was the end of an age for those who had peaked in the twenties and the beginning for such names as McCullers, Hersey *(Hiroshima)*, Stafford, Warren, Mailer, Jones, Salinger, Ellison, Malamud. Kazin is writing twenty years after 1945, noting ". . . the most violently accelerated living in human history":

> Twenty years in which the words Auschwitz and Hiroshima have become banalities of accusation, so that the continued refrain of book after book is our guilt. . . .You did it, you son of a bitch. You and me–we are guilty. It was the times. It is the age.[90]

In one of the periodical examination pieces cited above, Robert Coates discusses "The State of the Novel" in the *Yale Review.*[91] He finds the form at a new low, with few holiday listings and no Pulitzer award. Why?

> We live in what we are constantly being reminded is a 'scientific' era, the age of 'technics' and mechanization, of physics and chemistry and metallurgy. The result, as applied to literature, is that we tend to prize *facts* above *ideas*–and as a further consequence, to value any *book* about facts. . .far above any book about ideas, or in other words, a novel.[92]

Wallace Stegner thinks along the same lines:

> Novelists will not really 'discover' America until they cease to be dominated by the ideas of determinism derived from science and the post-Victorian climate of opinion.[93]

Frank O'Connor finds technology wrecking the novel which, in turn, produces a mechanical view of man, leaving us only one choice, to die or live. In *The Mirror in the Roadway*, 1957, he discusses James Joyce and dissociated metaphor, "bogged in technical devices." The trouble with *Ulysses*, he says, is that it reduces man himself to a metaphor,

> . . .a step that is openly taken in *Finnegans Wake*. . . .From Bloom, it is only a step to H.C.E., who is merely a metaphor in the mind of God and has no personal existence.
>
> Like the atom bomb, this can only result in the liquidation of humanity, and humanity has no choice but to retrace its steps and learn the business of living all over again.[94]

Just as Sypher sees the Machine as enemy of man as man and of man as artist, along with his art, so does O'Connor voice the not uncommon feeling that technology, technique, technics—whatever you wish to call it—is the enemy of man as man and man as novelist, along with the genre itself. Nor is the testimony all from the professionally humanist side. René Dubos, writing as a microbiologist viewing the human condition in his Pulitzer Prize-winning *So Human an Animal,* points out that

> The age of affluence, technological marvels, and medical miracles is paradoxically the age of chronic ailments of anxiety, and even of despair. Existentialist nausea has found its home in the most affluent and technologically advanced parts of the world. . . .The most poignant problem of modern life is probably man's feeling that life has lost significance.[95]

I have noted earlier, and have quoted observers to this effect, that it is not always *easy* to find specific evidence of the impact of post-Bomb technology on literature, but that there is no doubt that anyone of keen sensibility can detect a general mood of tragedy and nihilism in some contemporary poetry and many plays. And, as we have seen, there is ample specific evidence if the researcher will work to dig it up. In the case of the novel, as we have witnessed with Burroughs, the job is often made easier by a tendency in certain authors to run to the confession box. The Englishman, John Braine, is just such an author. In "People Kill People," the man who wrote the best-seller *Room at the Top* tells us of the problems he and his fellows carry with them:

> In order to write novels it is absolutely necessary for me to forget the existence of the H-Bomb. At least, I do as far as is possible push it to the back of my mind. Most of us in what I'll call the target countries have to do this simply to preserve our sanity. I don't think that my novels are unreal because the H-Bomb isn't much in evidence in the thoughts or conversation of the characters; in any case the consciousness of the thing is there. . . .
>
> I find it almost impossible to write about the H-Bomb. As far as that subject is concerned I can't separate my intel-

> ligence from my emotions. It frightens me too much. . . In fact, whenever I think about the H-Bomb, I can only give way to despair, which is to say that I can only stop thinking of it.[96]

Such fear is evidenced on our side of the Atlantic as well. At an *Esquire* conference in 1960, Martin Green tells us, Philip Roth, John Cheever, and James Baldwin, all "name" novelists, were discussing writing in America today. Baldwin said there is "no structure in American life today and there are no human beings." Cheever added that "the only possible position for a writer now is negative." Green reports further:

> At the end of the conference someone in the audience pointed out that all the speakers had presented themselves as victims. 'Exactly', they replied, delighted, 'we are all victims.'[97]

His comment is, "We can all recognize hysteria when we hear it." To recognize is one thing; the question then must be for humanist and technologist alike to ask themselves, "What should or can we do about it?" Irving Howe admits the hysteria but attacks the problem soberly and without pulling any punches. In "The Fiction of Anti-Utopia," he considers such sub-genre novels as *We, Brave New World,* and *1984*. He quotes Max Eastman, "I feel sometimes as though the whole modern world of capitalism and Communism and all were rushing toward some enormous, efficient, machine-made doom of the true values of life."[98] According to Howe, the anti-utopians feel that doom is here:

> [This is] the nightmare-vision of the anti-utopian novelists: that what men do and what they are become unrelated; that a world is appearing in which technique and value have been split apart, so that technique spins forward with a mad fecundity while value becomes debased to a mere slogan of the state. This kind of 'technicism,' Spengler has remarked, is frequently visible in a society that has lost its self-assurance.[99]

Among the ideas Howe brings up—"seminal" ideas in the jargon of today—are the following: "the need felt by modern man to drop the burden of freedom"; "the idea of the personal self seen as a cultural idea, a fact within history. . .susceptible to historical destruction"; and the problem first raised by Dostoevsky, who wondered if the satisfaction of material wants would quench the appetite for freedom and who noted "the misery of the human being who must bear his burden of independence against the contentment of the human creature at rest in his obedience." The fear of these novelists is documented on TV and advertising billboards which tell us in effect that "the secret of happiness and virtue is liking what you've got to like." No wonder De Tocqueville (as quoted by Howe) feared that

> a kind of virtuous materialism may ultimately be established in the world which would not corrupt but enervate the soul, and noiselessly unbend its springs of action.

From confessional of despair to hysteria to prophetic doom leaves the ultimate extrapolation up to the science fiction writers, who have for many years now enjoyed recognition on both sides of the tracks. Even the stuffy Modern Language Association now admits papers on science fiction at annual meetings. It is interesting to note, incidentally, that the number of titles in this one area of novel writing took a tremendous, and logical, leap upward in the annual record of *Reader's Guide* after August 1945.

Among the science fiction novels discussed in Mark Hillegas' *The Future as Nightmare*, three may serve to illustrate the ultimate in projection. After all, from Jules Verne on, many science fiction notions have come true. Why not some of these? Ray Bradbury's *Fahrenheit 451* treats of the "virtuous materialism" and enervation of soul about which De Tocqueville spoke so eloquently: under the threat of nuclear war Americans are absorbed in four-wall TV, radio, supersonic travel, and sports. "Life is reduced to the paste-pudding norm of a mass audience, for it serves the purpose of the government to keep people from thinking: Why learn anything save pressing buttons, pulling switches, fixing nuts and bolts?"[100] Kurt Vonnegut in *Player Piano* brings out the point met already in expository articles, that automation takes away the fun of working with one's hands and leaves one with nothing useful to do; furthermore, Machines have now begun to make decisions because they have become smarter than men. Finally, in *A Canticle for Leibowitz*, by Walter Miller, Jr., mankind has reverted to the Dark Ages following a nuclear war referred to as the "great Flame Deluge." After a long, painful comeback to another Renaissance, man then moves into a new technological era. Eventually nuclear weapons appear again. The world once more is destroyed, and the few survivors undertake "a new Exodus from Egypt under the auspices of a God who must surely be very weary of the race of Man." [101]

Anthony Burgess, author of many successful novels, has said in *The Novel Now*, 1967, that nuclear war is not the sole property of the anti-utopians or science fiction writers: "A good deal of ordinary non-visionary fiction has the shadow of the Bomb in it." [102] But Burgess feels that "few good novels came out of that real, historical war which ended with the blasting of Hiroshima and Nagasaki." The reasons apparently include, he thinks, the grimness of modern war, the different spirit of disillusioned youth of World War II as compared with that of Rupert Brooke's 1914 group, and the inclusion of civilians and their suffering from mass bombing. War has lost its glory, and there is no sense in recording the new style. (It is interesting once more to note the dominance of the Bomb image in British writers.) In discussing Muriel Spark's story *The Girls of Slender Means*, which is a symbolic novel about some strange characters who are members of a club, in the garden of which an unexploded bomb is said to lie buried, Burgess stresses the fact that no one takes notice. The bomb goes off–in 1945–destroying the club (human society). Our analyst comments, "And yet we can accept these strange people making up a world like our own, since the world that has dropped an atom bomb in its own garden cannot really be accounted sane." [103]

The final work to be introduced in this section is James Gindin's *Postwar British Fiction: New Accents and Atti-*

tudes, 1963.[104] The author has a good bit to say of the young British novelists with university background, people like John Wain again, Kingsley Amis, and Iris Murdoch, whose work is quite popular over here. These writers have tried to reintroduce the central character who tries to get ahead in society, but with a difference traced to modern loss of self-assurance. Although these stories come out as oddball comedy, they clearly show modern man floundering without a formula. Why? "Two world wars, the threat of the hydrogen bomb, and disillusion with the Marxist version of world brotherhood." Again, the Bomb. Of the examples given by Gindin, perhaps the best is Alan Sillitoe's highly successful novel, *Saturday Night and Sunday Morning.* First, let us have a passage from the book, later made into a hit film:

> Because it was no use saving your money year after year. A mug's game, since the value of it got less and less and in any case you never knew when the Yanks were going to do something daft like dropping the H-Bomb on Moscow. And if they did then you could say ta-ta to everybody, burn your football coupons and betting slips, and ring up Billy Graham. If you believe in God, which I don't he said to himself.[105]

And now Gindin comments:

> The possibility of the H-bomb creates a good deal of the uncertainty in the background of Sillitoe's characters, yet they do not tremble when they speak of it or join committees to advocate a sane nuclear policy. On the contrary, Arthur rather enjoys telling his farmer uncle about the possible effects of radiation. The H-bomb is less an appalling horror than further evidence of life's essential uncertainty, another vast and unpredictable possibility that must be lived with.
>
> Still, wages are good and no H-bomb has fallen yet. Thus Arthur is lucky and he recognizes it.

In the long run, however, the attitude of post-war British fiction and of a sizable minority of American novels is "the existential attitude, with all its qualifications and its comic incongruities, [which] seems appropriate for the civilized man caught in a civilization that, as he well realizes, may not last very much longer."[106]

The attitude of the contemporary novelist, then, says in a different way and a different medium, substantially what some poets of today, many dramatists and, for that matter, many anthropologists and biologists (notably René Dubos)[107] have said in their own ways: Things are indeed in the saddle, they are moving too fast, and man is finding it increasingly difficult to cope, being caught in a space-time dilemma. Let Stephen Spender, English poet and critic, have the last word here. In his essay, "The Struggle of the Modern," he says,

> The modern is acutely conscious of the contemporary scene, but he does not accept its values. To the modern, it seems that a world of unprecedented phenomena has today cut us off from the life of the past, and in doing so from traditional consciousness. At the same time it is of no use trying to get back into the past by ignoring the present. If we consider ourselves as belonging not just to our own particular moment in time but also to the past, then we must also be fully aware of our predicament which is that of past consciousness living in the present.[108]

Perhaps technology will tell us how to set our time machines?

6. . . .and "Other" Literature: Finale

I have said little about two other major types of literature: biography, a personal favorite which, however, does not lend itself comfortably to this present investigation; and the essay, which has fallen on evil days. The informal, pipe-and-slippers essay in the tradition of Lamb and Hazlitt has all but succumbed in our day to the pace and style of modern living; Christopher Morley and the early E.B. White of *The New Yorker* were about the last of their race. The formal essay has for the most part been replaced by a slick cousin, the article, with no great claim to style (one actually reads pretty much like another) and with its main purpose not classical charm but expository fact and opinion, to inform and persuade. The article fits somewhere in between two definitions of literature: one, the kind I have been talking about, which is creative and genuinely "literary"; the other, what may be called professional in the sense that a doctor may say to a colleague, "I have seen nothing in the literature to equal this new operating technique." The article is written adequately-to-well, by people who ordinarily do not write to live or who write part-time, and it has a valuable function, sometimes literary in the best sense and sometimes not. Often a bunch of articles becomes a book. Often, too, we run into something called simply a book, which contains chapters, each of which might have been an article, or which is really one overgrown essay. Of these, especially on science, technology, and the humanities or fine arts, Lewis Mumford recently wrote, "What a spate of books there has been since I wrote *Technics and Civilization!*"[109] I refer to such writings as those by Ashby, Barzun, Bronowski, Conant, Commoner, Ellul, Aldous Huxley, Lapp, Muller, Don Price, Rabi, Weinberg, Dubos, Holton, Krutch, and many others, who may be left respectively and respectfully in my bibliography. They need no help from me.

Since we began this treatment with a general frame, we may as well conclude with a matching piece. We left the poet Spender in Section 5 bemoaning "our predicament which is that of past consciousness living in the present." Quoting Mumford on this same breakdown of the continuities of history under the Machine *(New Yorker*, March 6, 1965), Leo Marx adds a comment which nicely sums up what I have said or quoted so far:

> . . .[Mumford's] is an attitude toward industrial society that permeates modern literature. It implies that the course of contemporary history is largely, perhaps irresistibly, determined by the course of technological development. And however much we may disapprove of this fatalistic idea, it does *seem* to be confirmed by the fact that when technical skill makes possible a flight to the moon, the building of an H-bomb or a supersonic jet, our society seems invariably to follow the lead of technology. Given a world dominated by such a machine, in any case, the pastoral impulse to withdraw (or to 'drop out,' in the idiom of today's radical youth), is an impulse to recapture a human situation as it might be imagined to exist beyond, or to have existed anterior to, our intricate technological order.[110]

André Malraux has recently taken up the same point. He is sure that "...the problem of youth is just one aspect of the most basic problem of our civilization—which is built around machines and neglects man. . . . Humanity, I believe, is conscious of this. . . .And so there is this extraordinary malaise, especially among the young."[111] René Dubos reaches Malraux's intuitive conclusion from scientific study:

> History confirms present-day observations in demonstrating that man can become adjusted, socially and biologically, to ways of life and environments that have hardly anything in common with those in which civilization emerged and evolved. He can survive, multiply, and create material wealth in an overcrowded, monotonous, and completely polluted environment, *provided he surrenders his individual rights, accepts certain forms of physical degradation, and does not mind emotional atrophy.*[112]

Such a surrender, of course, is unthinkable for the post-1945 artist and writer.

From Dubos' "atrophy" to Fromm's "apathy" is but a typological step. In *The Revolution of Hope: Toward a Humanized Technology*, the latter sketches the alternatives to apathy:

> The alternatives that face man in these apocalyptic times lead either to a mechanized society, in which man is caught like a helpless cog, and to eventual destruction by thermonuclear war, or to a renaissance of humanism and hope, a society that puts technique not first but second, as a servant of human well-being.[113]

This renaissance of humanism ought to come from the humanists, I suppose, but most of them are locked up tight in their ivory towers. It looks as if the challenge will have to be taken up by humanistic scientists—or so J. Bronowski would have it:

> Science as a humanistic discipline has to transmit and inspire this sense of uniqueness ('the unique and double creature: man, the social solitary'), and to found it on the order of nature and not on the primacy of man. . . .But that does not mean that it turns him into a beast or into a com-

puter. On the contrary, what makes the biological machinery of man so powerful is that it modifies his actions through his imagination: it makes him able to symbolize, to project himself into the consequences of his acts, to conceptualize his plans, and to weigh them one against another as a system of values. . . .

The humanist reality is that man is guided by values and that he creates them for himself. This is the hard discipline which it now falls to science to teach in a world that has lost the comfort of being sustained by any absolute purpose.[114]

Man, we may remind ourselves, was a symbol maker before he was a tool maker. The symbol makers in poetry, drama, and fiction, along with their brethren in painting and sculpture, devoid of the comfort of any absolute purpose, have been singing the same song, if Bronowski needs more documentation. The scientist and the artist meet before the great truths, and life and literature blend to a blurring point. A curious and fascinating example of this convergence, the realization of which lies at the heart of one culture as opposed to two, may be found in a recent prize-winning book by Robert J. Lifton called *Death in Life: Survivors of Hiroshima.*[115] In Chapter 10, Lifton discusses *A-Bomb Literature*, the diaries, memoirs, and poems of the Japanese who lived through the holocaust. There are poems which curse "bad America," and later "poetry of calm anger" which shift their emphasis to the world and general evil. There are poems of protest like Sankichi Toge's "August Sixth":

How could I ever forget that flash of light!
Treading upon shattered human brains. . .
heaps of schoolgirls lying like refuse

and elegies of reconciliation like one by Eisaku Yoneda, who regarded annihilation as part of a biospiritual continuum looking toward reconstruction. These direct testimonies mark the ends of the spectrum in literary work since the Bomb. And curiously, the feelings writers have shown, although they have only read about or seen pictures of Hiroshima and Nagasaki, are almost identical with feelings reported by the actual survivors. The degree of reaction since 1945 is more marked because psychologically our poets, novelists, and dramatists around the western world were there, so to speak. Compare what American, British, and continental writers have said above with these testimonial phrases gathered by Lifton for an article. "On Death and Death Symbolism: The Hiroshima Disaster":[116]

'immersion in death'
'death in life'
'helplessness and abandonment in the face of annihilation'
'invisible contamination'
'enduring taint'

The real victims felt that all previous notions of immortality and support had been swept away—immortality through descendants; through an after life; via literature, art, and one's inventions; and through eternal Nature. These are the notions by which man has lived. The synthetic (but real in another sense) victims, the writers, alike have felt the loss of belief, immortality, decency, reason for any but an existential philosophy, inspiration, individuality, and external Nature, and from the same cause. As Lifton says, if death gives meaning to life, what kind of life with this kind of death?

7. . . .With A Coda on Technology and Art

K.G. Pontus Hultén, in the Foreword to *The Machine,* a beautifully illustrated catalogue put out by the Museum of Modern Art in New York, 1968, gives a useful quick review of the relations between art and technology in the early part of this century. He traces the work of Tatlin in Russia, who in 1920 tried to fuse sculpture, architecture, and motors for a great memorial; of Moholy-Nagy and the Constructivists, who were optimistic about Machines; of Leger and Le Corbusier, Purists who found elegance in Machine forms. Then came a shift with the Surrealists, who by 1924 depicted Machines as enemies of Nature. The

1929 Depression intensified disillusion with technology, as seen in the lines from Chaplin's film, *The Dictator:* "We think too much and feel too little. More than machinery we need humanity."[117] But the world was headed in a different direction (1945):

> The bombs dropped on Hiroshima and Nagasaki were the most terrible shock that the world has ever received. Fear and horror sapped the faith in technology and the confidence in rational behavior that might have been expected to follow a long period of destruction.[118]

As Hultén tells it, artists then turned away from relating technology to life, and the New Constructivists concerned themselves only with formal problems. By the mid-fifties, however, Tinguely and others sought better relations with technology, feeling that misuse of it had been to blame. Pop artists tried to find a way out of alienation by reminding the public that mass products are, after all, the creation of human intellect. But there was still the lurking "notion that modern technology has an evolution of its own. . . uncontrollable and independent of human will."[119]

Modern and contemporary artists, then, have reacted pretty much the same to technology as have the writers discussed thus far; that is to say, they fear it much of the time and welcome it some of the time. They certainly cannot be tossed into one basket labeled "The Enemy." To begin with, Mumford sees a direct connection between today's art and the Machine:

> The only group that has understood the dehumanizing threats of the Invisible Machine are the *avant-garde* artists who have caricatured it by going to the opposite extreme of disorganization. Their calculated destructions and 'happenings' symbolize total decontrol: the rejection of order, continuity, design, significance and a total inversion of human values which turns criminals into saints and scrambled minds into sages. In such anti-art, the dissolution of our entire civilization into randomness and entropy is prophetically symbolized; . . .by a different route both (technicians and artists) seek or at least welcome the displacement and the eventual elimination of man.[120]

Wylie Sypher agrees in spirit and substance when he says that "the technological frame of mind evades the possibility of surprise, the hazard in art, the grace beyond the reach of rule, the magic hand of chance." Pop art, Burroughs' novels, Happenings, and computer poetry are thus explainable revolt: "Such art is reckless to the degree of insolence. . .justified only by recognizing that our society is even more colossally insolent in attempting to engineer human beings."[121]

Now, as we have seen, technology frees man from drudgery and sets up leisure and abundance. But the price, as George Kateb sees it, may rule out most heroism, endurance, many kinds of ingenuity, many hallowed attainments and disciplines, many consolations and inducements.[122] Among consolations and ingenuity, of course, is art; it may go to caricature, as Mumford suggests, or it may run scared, scared of loss of the good life. Kateb sees the good life involving the following ingredients in ascending order: laissez-faire, greatest amount of pleasure; play, in the sense of games, ornaments, masks, style, or virtuosity; crafts; political action; and the life of the mind. The artist would certainly welcome the first four, and some of the best would enroll under five and six. The plight of the artist, even more perhaps than the writer, is that he sees, feels, and expresses things that are missed by many citizens but has to pay the price by seeing some things he does not want to see, like, say, the Machine threat to the good life in art.

Where Siegfried Giedion welcomes the union of engineering and architecture and Gyorgy Kepes hymns an alliance between science and art in the new landscape[123] some modern artists and art critics go below the surface and often don't like what they see. Such a figure is Sir Herbert Read. In an article titled "Art and Life," he first discusses modern literature with its violence for the sake of violence, modern boredom and the fear of being alone, and the whole neurotic syndrome based on a sense of insecurity. As might be expected, Sir Herbert states,

> This universal neurosis has developed with the progress of technology. It is the neurosis of men whose chief expenditure of energy is to pull a lever or push a button, of men who have ceased to make things with their hands.[124]

Wylie Sypher has dredged up the French word "bricolage" for what Read is talking about; it means "a way of knowing reality by what one does with one's hands." The argument that technology has deprived men of the needed chance to work close to Nature with their hands is a favorite one among artists (and some anthropologists). Read, in talking about button pushing, is like Rice in *The Adding Machine:* he feels, again like Rousseau, that "intelligence develops in contact with things" and "the majority of the population has little or no sensuous contact with the soil, with animals, with the handling of wood, clay, or metal." Against the threat of the assembly line, Read advances a second common view among artists today, art as defense through creativity, and calls upon the educational process to bring the victory by selecting, in the words of Walter Gropius, "young people before they have surrendered to the conformity of the industrial community or withdrawn into ivory towers." Read and Gropius proposed to bridge the "gap between the rigid mentality of the business man and technologist and the imagination of the creative artist." Mumford, Kateb, Read, Gropius seem to be reminding us that there must be time to wonder, to imagine, to *not* understand lest we kill the magic which is life.

Wallace Stegner pokes at the empirical process which has "dictated the alloys in our burglar alarms. . .synthesized our insect sprays. . .and shaped the intellectual and physical life of nearly everyone" and finds technology to be the villain, not science: "There is probably a real quarrel between the arts and technology—what would Vermeer make of a General Electric kitchen?—but surely there is none between art and science. . . .Science cannot destroy art, the record of man acting." Stegner puts his finger on another point of contention between technology and art: Science is "open to exploitation as technology" leading to mass production of art and, more important, "What anyone who speaks for art must be prepared to assert is the validity of non-scientific experience and the seriousness of non-verifiable insight."[125] Charles Morris had said the same thing somewhat differently way back in 1939: to him art is like an icon, shows value, does not discuss it–it *is*.[126]

We have seen modern art thus far as revolt, as defense, as rescue, and for its own sake. Eugene Rabinowitch suggests another role, art as prophecy:

> . . .the feeling that all mankind is in one boat, to sink or swim together, is growing before the specter of a common nuclear catastrophe. Art may be the mirror and the barometer revealing and anticipating the struggle between the forces of atomism and fractionation of human conscience, human passions, and human society, and the creative forces of harmony and integration. Not through its choice of subjects, or its exaltation of this or that idea, but entirely in its own medium, and through its own means of revealing, liberating, and giving form and expression to the spiritual stirring of man, art could not only anticipate, but also assist mankind in finding an answer to the challenge posed by science.[127]

Rabinowitch, by the way, is a physicist, chemist, and editor of the *Bulletin of the Atomic Scientists.*

Consider the attitudes so far in this note on science, technology, and art: attack, warning, dismay, criticism, apartheid, and symbiosis. About the only reaction left out is that of enthusiastic support of technology in art, and that can be found too. In 1967 in Los Angeles, there was considerable publicity given to experiments calling for teams of engineers and sculptors to work together and learn from each other. The response was encouraging. In *Art in America* (winter 1968), Douglas Davis did a strikingly illustrated article called "Art and Technology–The New Combine."[128] The subhead read:

> Living as they do in a supertechnical society, American artists quite naturally turned to the products, processes, and imagery of science and industry. Some approach technology with traditional attitudes, others are using it to alter

> the very definition of art, but all who succumb to its fascination have responded with a new sense of exhilaration and discovery.

The article discusses hopping, tilting, horn playing, light-blinking objects of art in plastic, rubber, and aluminum. Charles Frazier, an artist, is quoted as finding "in the machine not only a means to an end but positive joy and exuberance." New effects gained by borrowing machines such as optical coating devices, by explosive forming done through welding and shaping high-density metals underwater by means of controlled blasts, by using new materials like styrofoam or acrylic plastic, and so on, join with computer music, computer graphics, and spectator-controlled light-and-sound forms to give a new meaning to an old relationship. The radical mutation is marked by: materials produced by recent technology, tools and methods borrowed from it, new imagery suggested by its tangible forms, and full partnership between artist and Machine in the creative process. The way to the future need not be gloomy. Says Davis,

> . . .ignorance is vanishing now. . .timidity receding, despite all the gloom engendered by prophets like Mumford, by facts like Hiroshima. We use technology now with exuberance. . .an exuberance impossible even a decade ago. . . .

There is always the danger, of course, that this kind of thing can go too far. In *Saturday Review,* February 8, 1969, Charlotte Willard reviews under the heading, "Presaging the Triumph of Egghead Automata," a new book by Jack Burnham called *Beyond Modern Sculpture* with its theme: "Today's sculpture is preparing man for his replacement by information-processing energy." Willard demurs, saying that "Love and hate are stronger than machines and electronic systems. . . .In the last analysis, the dematerialization of the planet depends not on technology but on our will to survive or our will to die." [129]

It is too early to tell what the final relationship will be between art and technology, and between literature and technology. We are only just beginning to see clear relationships between art and poetry, and between both of them and music. The whole process calls for new vision, interest, and work. A good deal will depend on educating properly the specialist people in both of the two cultures, not to make them one and the same necessarily, but to make them see where they resemble one another and why they should work together. A headline in the *New York Times* in November 1968, reads "Design and Technology Teamed";[130] the article concerns a new computer-controlled knitting machine which is so efficient that the designer can be more creative than before because of reduced problems and hence greater flexibility. A month later the *Times* carried this feature headline: "Sculpture at M.I.T. Merges the Arts with Science."[131] The article speaks of "a fast-moving new trend to merge the seemingly distant purposes of art and science," describes the new magnetic sculpture, and urges further exploration of the "interface between the two worlds. . .to achieve greater scientific insight and creativity." An artist named Takis is reported as saying, "We try to achieve spiritual collaboration between artist and scientist. Otherwise the technology is just a gadget. For me a scientist is a poet, a creator."

In conclusion, Hultén brings the two fields together from yet another point of view:

> The decisions that will shape our society in the future will have to be arrived at. . .through technology *But they must be based on the same criteria of respect and appreciation for human capacities, freedom, and responsibility that prevail in art.*[132]

Now if we can just get the technologists together with the poets and playwrights! If technology is the doctor, and artists and writers are quite articulate patients in a sick world, they can help by telling the doctor where it hurts, and how much. Any decent doctor schooled in method, technically trained, and experientially honed ought to be able to make a good prognosis if he can get the case history first and not have to work blind, as in an emergency, with an unconscious victim. He has the instruments to kill or cure. Is there any valid reason why the patient in this instance should not be listened to?

8. Conclusion: A Humanist Specializing in Literature Addresses a Reader-Colleague in the Field of Technology

Dear Friend:

We have just left the patient, "etherised upon a table," as Mr. Eliot would have it, and are now back where we began. If Sections 1-7 have been the Saturn booster, this is the tiny capsule hopefully in orbit (but not too far off the ground).

As a professor of English and former head of a department of humanities, I have been moving through the years at Carnegie Tech., Cal. Tech., U.S.C., Oak Ridge, Harvey Mudd, and elsewhere between the two worlds of the fine arts, on the one hand, and science, engineering, and technology, on the other. Blinded like Tiresias (in the present instance by the blast at Hiroshima), but with fear that I do not have his gift of prophecy, I am still groping my way between two worlds, trying to bring them closer together through understanding. My plight is a bit like that of Blanch in *King John:*

> Which is the side that I must go withal?
> I am with both: each army hath a hand;
> And in their rage, I having hold of both,
> They whirl asunder and dismember me. (III, ii, 327-30)

We began on the first page of Section 1 by entering the nearby woods. Now that we have had a walk down the woodland path, the question seems to be: What shall we do about the trees? Leave them for another look later? Perhaps plant some new species? Or maybe chop them down for timber or firewood?

I have my own answers about literature and technology and education, but with little evidence as yet, I may change my mind one day. There is no arrogance quite like that of the humanist who tries to tell the scientist what's wrong with science—unless it be the arrogance (myopia may be a fairer word) of the scientist who has all the answers. I will never forget Polykarp Kusch's remarks in the

Phi Beta Kappa *Key Reporter,* summer 1961, to the effect that every facet of our lives has been fashioned or modified by science or technology; he even went so far as to say that he could not "think of an important human need that cannot be satisfied by present scientific knowledge or by technology";[133] nor the reply by a young Brown commencement speaker named Fulton who said that if this were true, love and bravery and beauty no longer had any meaning. And I still feel a bit sad as I grow old and read what Rabi said recently (as quoted by Daniel Greenberg in his review of Klaw's *The New Brahmins, New York Times* Book Review, November 17, 1968):

> [What would happen if] we were to become a nation of poets and were taught in school. . .that every good citizen should write a poem? Some would be very good, and people would read and enjoy them, but what would anybody talk about? Only everyday things–love, sorrow, life, and death. If men want to go beyond these everyday things to a grand theme, they will find it only in science.

"*Only* everyday things!" Dear old Rabi! He has written some fine things on science and the humanities and the need for wisdom (See Bibliography). I spent a delightful evening with him once in Claremont, and he gently teased me about humanities in engineering education. But this! It reminds me of something Eric Larrabee once wrote in "Science, Poetry, and Politics."

> I am reminded [says Larabee] of a physicist with whom I once discussed the 'problem' of Anti-science; he said that it seemed to him perfectly natural that people should resent the scientist's superiority. There is a word for this, gentlemen, and the word is arrogance. It has nothing to do with science proper, it is not required by the needs of dedicated and impartial investigation; and it is certainly not sustainable on an evidential basis. It is an archaic prop to the ego, a social and psychological bad habit left over from the bad manners of nineteenth-century academic life, and fortunately it is already on the way out. But it is still one of the first and most unnerving aspects of science that many laymen encounter, and it has done incalculable harm.[134]

So I shall move carefully, humbly, often ignorantly, but honestly, for it is my deepest wish not to sound prejudiced or intolerant. And over and beyond this letter is the larger consideration: the necessity for humanities and technology to come together.

Some of the points raised in objections made by writers and artists against technology have been admitted in your professional literature. (One example would be Von Neumann's article, "Can We Survive Technology?" in *Fortune,* June 1955). But from my weak posture as a humanist acquainted with technology, your thinking is also committed to finding mechanisms, to proof by data, to only three-dimensional views, occasionally and admittedly to a sociological approach, but in short, to methodology, in a manner that would only serve to widen the gap between a proponent of further investigation into literature and art, and those already in full gallop in scientific research.

The humanist, including writer and artist, (and I am speaking now as part-time writer) looks for warmth, an anecdote here, an actual case history there, a little humor, a well-put one-dimensional opinion in any discussion of modern technology. He wants programs about people to sound as if human beings were alive and not statistics. Or numbers. But you can't do this and remain coldly objective, scientific, methodological–I can see this, and I don't know how literature and art can be handled that way. But I would still suggest that writers and artists have something valuable to offer those of you in charge of technology if you are not wholly committed to data. Let me quote a part of an editorial by Norman Cousins:

> The biggest single need in computer technology is not for improved circuitry, or enlarged capacity, or prolonged memory, but for better questions and better use of the answers. Without taking anything away from the technicians, we think it might be fruitful to effect some sort of junction between the computer technologist and the poet. A genuine purpose might be served by turning loose the wonders of the creative imagination on the kinds of problems being put to electronic tubes and transistors. The company of poets

> may enable the men who tend the machines to see a larger panorama of possibilities than technology alone may inspire. . .there may be a tendency to mistake data for wisdom. . . .The poet can remind us that man's greatest energy comes not from his dynamos but from his dreams. [135]

A little schmaltzy, maybe, but is it nonsense?

Let me ramble a bit. After all, this *is* a letter. Isn't there a place in a discussion of technology and society for other kinds of vision and feeling, especially when society is made up of people who on occasion "dig" poetry, painting, and music? The findings of literature can serve as a brake on the maddest plunges of technology (cf. Commoner's Project Starfish) and a boost to the imagination of the creative engineer. Again, how could any enterprise already committed to values not concede the importance of literature as a repository and record of contemporary attitudes and values?

Your bias shows and so does mine, but they are not incompatible. I just happen to feel that the poet and the novelist, the playwright and the artist, speak for the society you are dealing with, and that their voices should be listened to *along with* those of trained researchers. Every time the team approach has been tried, members of both sides have been amazed at the insights gained. The function of the poet historically has been, as Sidney said, "to teach and to delight." In an age of technology, he still can–if he is allowed to. His may be a one-dimensional, facile view (it often isn't), but it exists and has some influence and at least heuristic value. In these parlous times, we see people reprocessing slag because they are running out of ore; if you run out of ideas, try reexamining the "slag" of literature?

Is it or is it not a function of those causing technological change to consider the trauma of change? If it is, then aren't even weak or despairing voices useful in estimating procedures and conduct of future research? One doesn't have to pander to these writers, of course. One can understand the contempt of a strong man for the weak; we all get fed up with some of the whiners. But I also have compassion for them and would like to help them by dialogue, another high-fashion word these days. Speaking of dialogue, why don't you technology boys set up a public relations Secretariat for Non-Believers for information and clarification, along the lines of USIS? The images and stereotypes discussed in Chapter I are still being circulated.

People like Leo Marx, René Dubos, and the anthropologist, Jacquetta Hawkes, think man has a basic need to get back to Nature now and then; Marx even thinks urban planners should pay attention to this need. Don't these people deserve a closer look and an answer? Even if there is only "some" truth in what they say? To dismiss them because they don't follow the technological party line is to say with the kid who owns the football, "You guys are gonna play my way or you don't play at all." The resulting game isn't much fun for anybody.

The views of poets, dramatists, and novelists may be too uncritical or too partial to guide inquiry. You won't get a mechanism out of them! They are not trained sociologists. The sociological approach is a good one when you choose to see it, but is it the only source of evidence you will consider outside the laboratory or manufacturing plant? I keep thinking of Cousins' remark about mistaking data for wisdom. I just wonder sometimes in frivolous moments if technology couldn't once in a while organize knowledge for impractical purposes, with the help of artists? Could it subscribe occasionally, just to clear its coronaries, to an ethic of waste, to use Sypher's phrase, make people a little less worried, showcase itself better in Madison Avenue style, and simultaneously accept the tremendous duty of warning its customers of the possible dangers of its products? (I'm getting too evangelical).

Getting back to the original question about literature, art, and the future of technology, I would say that an overall value of consulting in more depth the materials sampled only in Sections 1–7 would be a non-method approach to non-scientific truth and point of view both for the worth of the ideas presented and to insure against the accusation of technical blindness on the part of your establishment.

Although literature and art are often rational, you would have to accept intuition (any good mathematician knows what an intuitive leap is!), feeling, and emotion—but it just might pay off. Some technologists might enjoy or benefit from a poet like Robert Graves, who had some fun with the M.I.T. faculty in 1963. He asked them what sustained them:

> What is your secret mystique? Have you technologists a mystique? So far I have found no evidence of finding any such, but only a sense of fate. 'We must go on and on.' Why? [136]

Graves found in technologists an outward community, a limitation to objective views, factual accuracy, a shutting off of emotions—in short, a forgetting to live and to be better as professionals and people. One indirect benefit from pursuing literature further might be a reminder that there are other things in life. Another might be valuable knowledge of the enemy, if there is to be war.

In Sections 1-7 the writers and artists have already presented their case for art and literature as defense, as bolster, as different viewpoint, as record, as useful tool, and so on. The same ambivalence seen in nineteenth-century writing still obtains, but with more evidence of fear than awe and respect, though they are there too. I would consider carefully the artist's role in a needed return to general culture in an age of specialization. Further, it would be good intellectual exercise for the technologists to find out what art and literature say about science and technology and then come up with their answers. Why go to literature and art for ideas obtainable in the public press? Answer: for depth of perception and style of presentation. Question: Is the inner horror of some poets about the Bomb or the Machine a fair index to the feelings of the inarticulate national body? The answer needs further study. Shouldn't we heed the artist as spokesman for the masses, who have no lobby like Science and Big Business? Some more questions that might be studied:

How important is it to learn the aesthetics of decision-making?

How really important is it for man to stay in literal touch of earth?

Is it really important to heed biologists' warnings of the limits of human adaptability sensed by writers too?

Is a person really better off for writing, painting, sculpting?

What can urban planners learn from poetry about pastoral background?

If plays are full of violence and filth, how far do they reveal public reaction against the cruelty of Hiroshima and a mechanistic universe?

Can technology and art together reveal new forms?

How far can technology widen the materials of poetry? the language?

Can, or should, literature and art assimilate the Machine?

Is methexis, active individual participation, in art really lost in an assembly-line, mass-production age, and is it that important anyhow?

How far, how measurably, can technology be blamed for Kafka, Camus, and the Theater of the Absurd? Or should it be thanked?

How to weigh the artist's charge that technology has depersonalized life against recent artists' joy with new materials furnished by technology?

How valid is Lindsay's point (Bibliography) that the humanist's most damning indictment of modern technology is the role he thinks it plays in forcing conformity in the social habits of people?

How do we separate or sort out Bush's opinion (Bibliography) that "All modern poetry has been conditioned by science, even those areas that seem farthest removed from it", and J. Isaacs' flat denial of this in *Background of Modern Poetry* (Dutton, 1952)?

The poets Robert Lowell, Theodore Roethke, Anne Sexton, and Sylvia Plath, all contemporaries, either committed suicide or spent time in asylums, appalled by modern life. How far is technology responsible?

You see, I get older and scare more easily than heretofore. When I read that technology enlarges choices, I also know that choices from new outside-determined options are guided choices, not free. And if your people talk, as some do, of fitting religious belief to the technological age (not the other way around, being realistic), then you can talk one day of tailoring other beliefs, and the result will be the intellectual dictatorship of the Machine or of the bureaucrats running it.

Let me go back to poetry once more, to the notion that man lives in part on myth, that art and literature supply this need, and that modern technology has killed myth, or wonder, or illusion, if you will. (It remains to be seen how much the recent trip to the Moon does for wonder—promote it or end it, since fiction has come true). Here is Paul Ginestier quoting Paul Valéry and adding his own comment. The passage sums up the feel of my letter, and the last sentence the thrust of my argument for more understanding between those in literature and art and those in technology:

> 'Myths (says Valéry) are the souls of our actions and our loves. We can act only by moving toward a phantom. We can love only that which we ourselves create.' This subtle definition (says Ginestier) explains the present popularity of myths: they associate themselves with the anti-intellectual movement which characterizes our era. Overwhelmed by the immense technical progress due to scientific reason and the methodical application of the intelligence, we seek to recover our psychic equilibrium and, by a common process of compensation, we have a tendency to magnify all that seems to escape the intellect. The more man becomes civilized in practice, the more he wants to return to the primitive in theory. The cults of nature and camping correspond to our taste for the irrational and to the attraction exercised by the most mysterious sciences. Thus the present popularity of the myth corresponds to a collective need rather than to a romantic infatuation. Sociologists would be wrong to neglect the study of these tendencies, for they constitute an important symptom of the sickness of our times (p. 10).[137]

Finally if you are still with me, if, as I believe, art and literature (and let's throw in music) record the best and worst moments of, and create the finest expression of, man as animal, hero, clown, and sign and symbol maker, I also have to point out to the tool makers that contemporary art and literature record, more sharply and clearly than before the Bomb, man's feeling that he is seeing his labors and his soul—even his existence—threatened by technology. If, as Gilbert Highet has said, the difference between a barbarian and a civilized man is that the latter enjoys the life of the mind (I would add heart), or, as Philip Rieff sees it, that the barbarian has nothing to remember, I would suggest that any methods of recording and preserving can never be studied too much. If representative artists talk of defense against, and some poets of retreat from, the Machine, maybe the Machine men should publicly convince the former that they are wrong—if they can. Technologists can hardly get away with refusing to debate because they insist on limiting the rules—this would convince nobody, laymen included. Further investigation of the life of the individual, of values in a technological world, must surely consider the ways of thinking and seeing of the artist and the writer, weigh their ideas for their potential value, probe areas where technology and fine arts can work in symbiotic relationship, and finally restore human meaning to the statistics and tables already set out as the distillate of research. I shall come back to this in Chapter IV. But first, let us all examine together what can be done about bridging the gap between two cultures in just one area, the education of engineers. After that, we will move into wider considerations involving other ways of getting together and involving any thinking, concerned individual, layman or professional.

Yours sincerely,

The Author

Chapter III

One Kind of Bridge: Humanities and Social Sciences In Engineering Education

"There is only one way out of this; it is, of course, by re-thinking our education."
–C.P. Snow

1. The Recent Past *

If there were world enough and time, we might review the history of technology as background for a discussion of modern engineering curricular practices in the areas of the humanities and social sciences as one example of an educational attempt to bridge the two cultures. It would be fun to discuss Plato's view, mentioned in Chapter I, that the engineer was a worthy man but that you would not want your daughter to marry one; or view with pride the honors heaped on the ancient *architectus*, his counterpart. The spectacle of Hamlet's unfortunate military engineer blown up by his own land mine (also mentioned in the previous chapter) might then lead to a reminder that West Point was the first engineering school in the United States, and from there to a study of the role of engineering in opening up the Old West. Somewhere along the way we would run into the aloofness held by people in the arts and letters area toward scientists and engineers, particularly, as we have seen, in nineteenth-century England, where it was not considered quite gentlemanly to soil one's hands in construction work or in the laboratory. From the debates of Arnold and Huxley to the memoirs of Herbert Hoover we would find material enough for our own debates on the character and status of the engineer. And in our own day, when the engineer can see his creatures blast into orbit or be called upon himself to participate in crucial political

* Portions of this chapter have been adapted by permission from U.C.L.A. Report EDP 3-68, May 1968, by W.H. Davenport and J.P. Frankel.

decisions, there would be data enough to round out a thorough study as a background for the poor planners who must ask themselves, "What kind of a man is this engineer? What does he need? Where is he headed? What can we do for him?" But since there isn't that much time, let's be content with raising and dropping these historical temptations, and press on.

When engineering educators ask what kind of man they want to produce for what kind of world, they are, of course, first interested in basic professional qualifications. There is nothing to do here with this matter except to assume for the moment that the technical curriculum is in good hands, even though they may be clenched fists at times. In the first part of this century we would not have had to go much further than that assumption. Characteristically, the only non-technical considerations would have been limited to the grudging admission that an engineering student could use a little English (especially in report writing) and probably a little economics. This was a far cry from the present situation in which we find the role of the engineering school on a university campus including the offering of courses for liberal arts majors and engineering executives out in the field, as well as requiring a goodly portion of humanities and social sciences for their own undergraduate students.

With the acceleration of technological change following two world wars, it began to be apparent that an engineer could not succeed as a specialist alone. (A good many people had known this for some time, but the principle of cultural lag was in full force.) The fledgling student would have to move about socially, be a good citizen, eventually act a supporting role as normal husband, and learn how to get along with all kinds of people in all kinds of job circumstances. A semester of English and a semester of economics could hardly provide all this.

At M.I.T. and Cal. Tech., to cite two outstanding examples, in the early forties literature, history, psychology, and philosophy began to poke their way tentatively into the science and engineering curricula. The American

Society for Engineering Education (ASEE) and kindred organizations began to allow space for humanistic studies in their annual meeting programs. Committees were organized, field teams were sent out, reports were printed, and college catalogs got fatter. The Grinter Report, the Gullette Report, the Hammond Report, and most recently the Olmsted Report[1] made it abundantly clear that an engineer could not live on bread alone, but that he would have to understand the interplay between technology and society; learn the processes of decision-making; acquaint himself with the best that had been thought, said, painted, and played; and, in general, strive to become a real person and a thorough professional with regard for others and a sense of responsibility. No small order, that. The converts went into action, fighting for courses and units, struggling against inertia, developing rationales, adding faculty, staving off diehards, exchanging model curricula involving "requireds" and "electives," and, in general, manfully trying to make a better man of the engineer and a better engineer of the man. If, at this point, the thin veneer of culture seemed to provide little else but material for cocktail party talk, no one worried too much. Few had been at the game long enough to know what to worry about. Aim for the mystic goal of twenty percent humanities and social sciences—and fire!

The Liberal Studies Division (its name has changed at least three times) of the ASEE offered annual papers, and curriculum committees began to experiment locally. People began to talk about "humanities stems" and minors and guided choice and relevance, topics still being argued. Over all was a sense of purpose and dedication, but programs varied widely and deeply, and matters clearly had not been thought through. Through the forties and into the early fifties there were changes and counter-changes; at least the problems were beginning to sort themselves out (and they will be outlined in the next section). We were still a long way, however, from the time (1968) when Cal. Tech. could announce that they would add *ten* social scientists in one appointment year or U.C.L.A. would offer a course in engineering for non-engineers (tit for tat) or Harvey Mudd College would try a pilot course in *Man, Science, and Society* at the request of student customers themselves. The notion still persisted that one course per semester in non-technical subjects, if one could squeeze that many into the whole structure, would accomplish the miracle. Even though advising, then (and now) was haphazard, and strange things were happening in the land (like counting R.O.T.C. for humanities credit because part of the course included the history of war), there was great initial enthusiasm; and, for a while, peace descended over the colleges, and a too early and too easy sense of duty performed allowed the educators to turn their attention to more pressing problems.

It would be nice to say that from the forties through the fifties (when some engineering schools took a second look and began readjustments in the light of experience) and into the sixties the march of humanities and social sciences in engineering education was ever onward and upward: more units, better courses, more support, great innovation. Many of us took this for granted. With three or four exceptions involving that many outstanding institutions, however, most of us were wrong. While some schools have progressed, many have continued in a rut, and some have lost ground. This is all a part of the ever-present problem to be discussed in the next section as a springboard for diving into a sea of proposals in which engineering education may well have to swim or sink.

In the spring of 1968, several review teams of engineers and liberal arts people went out in the field, visiting colleges of engineering and schools of engineering in large universities. Their purpose? To reassess the findings of the well-known ASEE report, *General Education in Engineering,* which had appeared twelve years previously, and which has served as bible and almanac of information for any faculties interested in what the boys were doing at Case or Carnegie, Stevens or Newark, Lehigh or Clarkson. The final report of this project, directed by Professor Sterling Olmsted at Rensselaer and supported by the Carnegie

Corporation, indicates that all is not for the best in the best of all possible curricular experiments. All schools do not have that mystic twenty percent. There is apathy here, grumbling among some engineering faculty there. Serious replies to project questionnaires indicate that the purpose of humanities and social sciences in engineering education should be utilitarian (report writing); cultural (knowledge); developmental (personal tolerance, imagination, values); and contextual (engineer as agent of social change). So far, so good. But on many campuses little is apparently being done to implement the expressed purposes. One reporter, in fact, noted in many locations apparently less discussion, cooperation, and communication among humanists, social scientists, and engineers than had been observable fifteen years ago. Curricula too often seem to present a series of unrelated courses. The report asks for more understanding of change, of historical perspective on man and values, of social process and the role of technology, of continuing education; in addition, it stresses flexibility, creativity, tolerance for ethical and aesthetic values, and so on. Offhand, it looks as if we are going to try to produce Solomons in quantity, in order to keep up with modern technological change, the fastest acceleration of its kind since man began keeping lasting records.

2. The Ever-Present

Before we try to see where we may be going tomorrow, it might be worthwhile to have a look at where we are, and where, in some details of thinking at least, we may be bogged down forever. If the present picture of humanities in some engineering and technological schools is a stagnant one, the cloudiness may be due to lack of current, too many academic dams, against which the brush and fallen trees have piled up, and similar barriers to free-flowing progress. With the highest ideals and the greatest faith, anticipation can choke on red tape and the details, details of modern organization man.

At first the poor administrators who are following a trend, they think, have to showcase their wares. Shall the students buy humanities one at a time, or in packages with the proper amount of social sciences? What kind of packages? Would free choice be better? Guided choice? And should these culture subjects be limited to the lower division, upper division, or no division? What about units? What proportion of the four-year course can be added to or taken away from the existing curriculum? It is difficult, if not impossible, to find a professor who thinks his course is expendable. It is difficult, if not impossible, to really persuade a student that he needs more non-major electives, especially when he may have a built-in prejudice against "all that long-haired stuff" and those "oddballs who are always trying to psych you out."

In addition to the "bugs" that crop up in any going university Machine (size of class, room space, lecture versus recitation, amount of homework, to TV or not TV) there are new "bugs" in a Mark I or Mark II humanities program, and it takes time, sweat, and argument to find the right immunizing shot or lethal spray. In the departmental and administrative conflicts, many a hardened faculty campaigner has been rendered *hors de combat.* As he is carried off the field, glassy-eyed, he can only with difficulty close his ears to cries of "History-oriented?" or "Theme-centered?"

One of the difficulties always with us concerns advisement. We want to get breadth and depth, dig postholes and fire shotguns at the same time, and it won't work very well. Not with one course per semester, certainly. To be sure, M.I.T., Cal. Tech., and Harvey Mudd (which got in late–1957–and could start clean) have time and space for twenty-eight to thirty-five percent non-technical courses, but most other schools for various reasons have trouble getting twenty percent, and many have not shoehorned their way into that figure yet. Few colleges can afford full-time humanities advisers; as usual the poor professor has to "double"–he may or may not have enthusiasm, interest in students, time, or experience. Even if he does, how can he, in cooperation with his colleagues, come to a considered decision on a basic problem which still stumps professional educators: free choice by the student under supervision ("cafeteria" approach), some choice and some "requireds," or all requireds ("chow line")?

If and when the foregoing issues have been ironed out, we then consider certain local problems, which vary with the institution. At U.C.L.A., for example, I found that many engineers took psychology but few took literature. The student explanation was a dual one: the psychology building was just next door while English literature was almost a mile away; furthermore, in the latter course, the non-major was thrown in with liberal arts majors, and while the experience should have been good for him, he was naturally afraid for his grade average. Now, if cultural choice is to be determined by relative propinquity of the shrine, however understandably, where are we? One staff member at a Cal. State campus told me recently about an extension of this choice principle which threatens the whole educational system and would knock the theories of John Dewey into the proverbial cocked hat: apparently many of this new school's students picked it because it was near the beach, and, more important, it had ample parking space! More seriously, and in varying degrees, we must consider the intellectual climate of any one campus; engineering faculty attitudes toward humanities in general (a chance remark in class may effectively sabotage a program); liberal arts faculty attitudes toward humanities and social sciences in engineering (both in principle, because the dirty phrase "applied humanities" may suggest itself, and in practice, because they may be asked to donate lectures on the other side of the tracks, so to speak); and administrative attitudes on such subjects as the relative advisability of a separate staff in humanities in the engineering school, all-engineer, all-non-engineer, or half and half—as opposed to one drawn part-time from ivory towers in the liberal arts sector of the campus. These are the things bad dreams are made of.

At one extreme, as Martin Green points out,

> Hostility to science or to 'an age dominated by science,' cannot any way serve the cause of the humanities, or the

> intelligence of the students to whom it is expounded. It is a form of stupidity, which teachers of literature everywhere should be actively stamping out. [2]

In the other, more enlightened, direction one finds the philosophy of such people as the architect Walter Gropius, who proposed in his Bauhaus School to keep students from either surrendering to industrial conformity or hiding in ivory towers by educating them "to bridge the gap between the rigid mentality of the technologist and the imagination of the creative artist."[3] In the same vein the poet Archibald MacLeish pleads for the production of truly educated men as he traces how the pre-Hiroshima notion that technology would serve human needs has gradually given way to the idea that it *will* do what it *can* do. How else, he asks, can we manage the technology which has evolved post-Hiroshima?[4]

It is becoming more and more clear that in the Technological Age management and control are at the heart of "re-thinking our education," to return to Snow's phrase; as we have seen, this involves refraining from hostile statements by either humanist or engineer, managing to steer a course between conformity and withdrawal, and using informed restraints on aimless production. Unfortunately, as Admiral Rickover pointed out in an address in February 1969, "today there is no absolute requirement that an engineer must be a liberally educated man, nor has engineering adopted the kind of ethical code that governs the older professions of medicine and law."[5] The tremendous drive toward professional achievement all too often works against a balanced educational program, unfortunately. And while men, as Kenneth Keniston says, "have a deep need for psychic integration," it is often forgotten in the mad race to succeed, to fit modern society's image: "The ideal of personal wholeness, of the harmonious integration of passion, conscience, and ego, is systematically undermined by our social definition of virtue."[6] Little wonder that we have dropouts, alienation, and student protests! Part of the trouble stems from modern man's face-off with the Machine as symbol of power, as we have seen in the preceding chapter. Lewis Mumford has rightly indicated the good and bad of the Machine and the need for education in order to know balance and control. Failure to educate properly, he writes, lets emotional drive drift: "it explodes in nihilistic revolt as it frets against the sterilities of its surroundings."[7] This is, of course, a thrust in the direction of argument for a more generalized culture, the topic of the following chapter. "The world is too much with us," as Wordsworth wrote at the beginning of the last century, and college curriculum planners are hard put to it, indeed, to attack even the broad and general aspects of the ideas presented by such astute observers as the half-dozen gentlemen just quoted in this digression.

Specific housekeeping details, with their particular application to humanities in engineering, admittedly a byproduct or ancillary enterprise, inevitably center around such points as the difficulty of courses offered, the value of interdisciplinary procedures, the possibility of self-study programs or directed reading for the best candidates (and how do you pick *them*?), pass-fail grading for non-major subjects, and the like. Whether these matters are decided first or after the broader topics such as goals and philosophy have been discussed, they are a challenge to any educator. Furthermore, just when everything appears neat and tidy, a new dean may arrive, or the students may stage a protest march, and back you go to the drawing board. Many an interested layman simply has no conception of such details.

If programs have foundered or stood still or somehow disappointed the believers here and there, it may not have been for reasons of load or types of courses or advisory systems. All too often more significant matters get overlooked in the debates over minutiae. Take the matter of goals. What kind of engineer are we planning to turn out? Are we interested in basic knowledge and adjustability in graduate school or industry, or are we going to formula-feed right here at home? These are some of the questions that must be worried over. It makes a great deal of difference to a humanities program, for example, whether the engineering faculty, for the most part, thinks of engineering as a job or a profession, and of an engineer as a money-maker or a do-gooder or a decent combination. When we start to talk about purpose and ethics and rewards that are not monetary, some old-timers may spit and growl.

The serious questions must cover theme and relevance. Should we move over Socrates in favor of Oppenheimer, as someone has recently suggested? What about listening to the students now and then? Without resorting to the lowest forms of permissiveness, the business can really learn something from the customers. One real stumper, for example, as a policy matter is finding out the best way or any good way to explain to fact gatherers, experimenters, and skill acquirers living in a quantitative academic world just what the place is in the world system for contemplation, meaning of meaning, depth of personality, quality, and what Matthew Arnold called "the need in man for beauty and the need in man for conduct."

Before closing out the ever-present problems in setting up and maintaining criteria for humanities programs in science and engineering education, and before leaping into the future, we must not ignore the difficulties of the flesh-and-blood human beings who work in this tangle of inanimate detail which we have been considering. A viable program depends on more than catalog requirements, equipment, and theory. A good program takes money. A good program needs first-rate faculty, and if we are talking about a humanities department in an engineering school, there are difficulties in attracting good people because they will not be teaching majors in their fields and because some of them suffer from the old notion that they will lose status outside of a liberal arts institution. If we are talking about retraining engineers to run their own humanities courses, we may find candidates in short supply. In an age of specialization the graduate schools rarely equip true generalists; even the last word is in ill repute among the purists. So there are problems of personnel.

We have indeed improved things from the days of English and economics. We offer more courses. We insist on the need for humanities and social science. We talk about continuing education and enlightened graduate curricula. Some schools have done a first-rate job. But, as the 1968 survey seems to indicate, others are just going through the motions. It has certainly become apparent here and there that you don't automatically produce a balanced, educated man by adding a few courses to the basic engineering or science requirements. As Lynn White has pointed out, it is not wrong or idle to give the students Picasso and Plato, but something else that's needed is missing in our relatively superficial programs: an awareness of a new atmosphere around all of us, a "recognition that science and technology are integral to mankind's adventure—the achievement of a unity of human knowledge and experience such as no earlier age has ever conceived." [8]

3. The Possible Future

Whatever we have been doing in humanities in technological education is not enough apparently. Not that all previous effort has gone down the drain—far from it. We are learning on the job. Programs developed across the country in such institutions as M.I.T., Cal. Tech., and Carnegie-Mellon have produced some positive results and a few feelings of frustration, but they have not been in vain. If properly regarded, they are products of a first stage in development from which we can learn, if we will, how to proceed. For, make no mistake, we have reached a turning point. We must decide whether the humanities-social sciences department or faculty is going to continue as a glorified service station, whether we must think only in terms of departments, divisions, and colleges within the university, or whether to take the larger view—not merely a dream but a practical possibility. Too, unless we are prepared to shuck the little boxes of credits, electives, ratios, and much of the academic impedimenta already discussed, we may be doomed to standing still at best, and the phrase "the plight of the humanities" will be accepted docilely with no show of irritation or defense.

If we review, in most instances we must confess that we have simply taken a few courses from the liberal arts curriculum and grafted them on an engineering curriculum, sat back hoping they would "take," and waited for a hybrid miracle. The result in some instances is akin to the TV commercial in which a man tries to stick a special charcoal filter from one brand of cigarette on to another; the whole thing clings for a minute and then falls apart. To put it another way, the thinking, useful as it may be, behind many curricula still going today is not unlike that which made the first automobiles look like buggies with a motor where the horse used to be. The idea was a logical application which quickly went out of date. Today we need new models in humanities education as well. But, as observed, first we have to decide what kind of engineer we want to produce. We ought to think one day of crossing over or obliterating departmental lines and talking in terms of cooperative single enterprise working to turn out a modern product to meet modern needs, an enterprise with real autonomy backed by money and respect. Today a student is expected to take a little English, art, sociology, history, or music–to mention a few subjects typically touched upon–and somehow to integrate and justify these components for himself. If he has a whole year in one area (two semester courses), we talk of concentration, sit back, and congratulate ourselves that Joe Doakes is not going to get by with eight semesters of eight different beginning courses (not a farfetched notion in some existing procedures). In effect, we give a man a few separate shots in both arms and hope that they will all merge somehow in a healthy system. He, on the other hand, talks of "passing off" language requirements, only rarely of "working in" the same. Is it any wonder that typically he doesn't see the light dawn until he is ten or fifteen years out, probably in some kind of management position? *Then* he feels something missing and asks why he didn't get more humanities when he was an undergraduate and what he can do about it now, as the recent studies of Purdue alumni by Perrucci and LeBold abundantly and convincingly show.[9]

Two years ago I produced the following:

> Our story cannot end with analyses of humanist and engineering attitudes and the record of past achievement. There remains the ever changing present and many tomorrows to consider. The historical debate on the impact of technology continues. The pros and cons of automation will long be heard in the land. And more and more the man in the street will become involved in changing patterns of living regulated directly or indirectly by scientists and engineers. Even the latter need reeducating, so fast is the world moving these days. When M.I.T. announces that its Ph.D. degree is out of date within three to six years one can begin to appreciate the shift from cookbook technology to training in basics plus conditioning for sudden adaptations and new orientations.
>
> Just as the history of engineering has shown triumphs and failures, the record of the present indicates possibilities of glory but also heavy responsibilities for the newly-trained graduate; he may one day build highways to the stars, but at the moment he has to do something about automobile safety, transport snarls, smog, pollution of crops, water, and natural habitat–all by-products of technology blinded or corrupted by human use of it. The future engineer, then, will have to repair the mistakes of others, keep up with the space age, tackle the food-population question, harness new sources of energy.
>
> It becomes clearer and clearer that the engineer's responsibilities do not stop with engineering. More and more he will find himself making awesome political decisions or becoming enmeshed in the nation's business. Indeed, in the world of tomorrow, scientist, engineer, and intelligent layman must for various and expedient reasons–practical, aesthetic, human, and humane–once more, like Francis Bacon, take all knowledge for their province. Technological know-how tempered by human understanding will dominate the scene in the foreseeable future. [10]

A voice from the sidelines asks, "All of this on 24 units of humanities and social sciences?"

The "scene" mentioned in the last sentence of the quotation includes the academic stage, of course. The plot, characters, and direction may change from day to day, but the show must go on, starring humanities, social sciences, and upstage in the spotlight, science, technology, or engineering. If the work of the recent past has done nothing else, it has through the record of its trials and errors, knowingly or not, indicated new approaches, particularly to engineering education. Distant signposts mention awareness of the price and threat of technology, training for administrative posts, importance of ethics, inculcating a sense of responsibility, growing up to decision-making, adaptation to constant change, recognizing and adjusting to value systems, recognizing qualitative as well as quantitative judgments, the problem of choice, and thinking through to the side-effects possible in an apparently well-finished operation. It is to be hoped fervently that planners and administrators read these signs clearly and pick their way carefully. They have a different look from fluid mechanics, strength of materials, English, and economics, don't they? And yet the latter are still basic, fundamental, and irreplaceable; without them the new phrases are but hollow sounds, meaningless, and unrealizable.

Professor Lynn White of the History Department at U.C.L.A. has pointed out that the first mark of a professional is knowledge of the history of his field. We might begin here as we look to the future. It would behoove engineering students to follow this line at present by taking a course in the history of technology if it is offered by the history department on their particular campus. (There is no real reason why one day a historian of technology might not be appointed to the faculty of the engineering school; he could also direct a basic freshman humanities course.) Whatever the set-up, in an age which gives birth to a quarterly called *Technology and Culture* (published at Case-Western Reserve) and an endowed Program on Technology and Society at Harvard, the findings of various past faculty committees strongly imply that engineering students should be aware of their traditions and their future role, both professional and non-professional, in the great world outside. (The same applies to science majors and courses in the history of science.)

Performance courses in fine arts, specialty electives like modern fiction, seminars in philosophy, and the like must at present continue to be offered by the staff in arts and letters until such time as graduate schools modify their offerings and departmental lines break down, a possibility which will be discussed a bit further on. The main point, whatever machinery is used, is to try through old and new approaches to give the student new vision, to continue to supply factual knowledge but to work also on developing in him such things as tolerance of ambiguity and a realization of similarities within apparent dissimilarities, so that the mythic power of the phrase the "two cultures" may be exploded loudly and finally. The engineering student will hopefully make it his part-time duty to "technologize" his liberal arts friends, his family, and off-campus associates so that all may better understand the age in which they live. As for the student himself, he should eventually be a better man and a better professional as a result of viable humanistic and technical training programs, both of which should continue throughout the four- or five-year period and after graduation; studies have shown an increasing maturity in upper-division students and a tremendous widening of horizons in graduates ten or more years out, as already observed. As Professor Leo Marx of Amherst pointed out in a lecture given at San Fernando Valley State in February 1968, student revolt, so common today, is really humanistically motivated and can be capitalized upon. Both sides are making mistakes: the students in thinking that singing about love and throwing their bodies on or in front of the Machine will furnish the answer, and the faculties in merely deploring and failing to turn youthful energies into purposeful activities. If faculty, said Marx, will work with these young people, provide historical perspective for them to show that man has been a symbol maker first (in art and literature) and later a tool maker, and restore the active, participating symbol-making process together with them, the result could be exciting and rewarding. Despair gets nowhere, Marx pointed out; the only alternative is nihilism. To be sure, the engineering student is characteristically less emotional, more stable than the typical demonstrator. But the notion of faculty and student working together in writing, speaking, painting, drama, and other forms of active doing and self-expression (not merely passively listening to theory and lectures and regurgitating facts on quizzes) is one that could and should carry over well into humanities education in engineering and science. And the notion of the historical relationship of man and symbol to man and tool, properly treated, could serve as the cornerstone of new building in the direction of an engineer's self-adjustment to environment. This is social science in action.

One destination of any good program involving science and the arts or engineering and the humanities is or should be unity, understanding, and cooperative labor for the general welfare, over and beyond the very apparent benefits, material and spiritual, to each individual student. J. Bronowski, physicist and poet, puts it this way in *Science and Human Values:*

> Whether our work is art or science or the daily work of society, it is only the form in which we explore our experience which is different; the need to explore remains the same.

He goes on to point out that we "remake nature by the act of discovery in the poem or the theorem."[11] This view of a larger unity and vision of common ground between opposing ideas or, for that matter, cultures, should be a goal in modern education across the board if it is to measure up to the demands of a fast-changing world. While most of us are neither poets nor mystics, we can hardly fail to sense the importance of ideals to work toward and goals to aim at; it is the purpose of this chapter to make a modest start on an outline of possible ways of viewing.

When we begin to take an ideal position and try to answer the question "What kind of engineer do we want to produce, and how much can we do about it?" some home research needs to be done, plus scouting around to see what is in the air. Finally, of course, we have to make some value judgments, stand up and be counted, and wait and see.

As a beginning, let's see what M.I.T., for example, asks high school deans and counselors to consider when recommending their best graduates for admission. A recent printed form lists the following: maturity, self-confidence, personal force, leadership, commitment, perseverance, desire to excel, creativity, imagination, originality, reliability, and willingness to accept challenge and responsibility. May we not speculate that these qualities should be maintained and developed throughout college and in professional life? If so, and we allow for considerable overlap in the above categories, we should try to work into the humanities program lectures, courses, and seminars which will stimulate intellectual growth, capitalize on the creative urge (which is remarkably similar in engineering and the arts),[12] and offer training in the techniques of problem solving. Another bit of evidence from a series of studies at the same outstanding institution cited above lists the following as desirable student goals: mastery of environment, ability to tolerate ambiguity, and willingness to take intellectual risks, the latter especially being something of a non-habitual attribute in the traditional engineering student, apparently. Following this line a bit further, it may again be noted that three-fourths of all engineering graduates these days, after they are about fifteen years out, find themselves directly or indirectly involved in management. Shall we then be bold enough to talk of applied humanities and social sciences and set up experimental courses in engineering enterprise, interdisciplinary in nature, like one designed by Professor Arnold Ruskin at Harvey Mudd? The latter in its broadest outline would combine elements of marketing, accounting, communications, decision-making, motivation, self-awareness, and personal growth, and would be run by a team of four men from four different disciplines. If the mathematician G.H. Hardy, in his *Apology* (recently reissued with a new foreword by C.P. Snow),[13] can talk convincingly about the place of applied and pure mathematics, it is conceivable that modern humanists can be convinced that there is a place for both pure and applied humanities, the former for personal delight and growth and the latter for use and benefit to the modern world, which needs to call on all kinds of attitudes and brains these days.

In another direction it is apparent that technological change is putting pressure on the schools and they are going to have to change with change. A recent book, *Dialogue on Technology* (Bobbs-Merrill, 1967), the product of many hands, indicates trends: one article, by C.R. De Carlo, entitled "Educational Technology and Value Systems" points out that "The technical act must exist *within* the fabric of the larger and more humane institutions" and that educated people must move "into a world of work in which daily activity will depend much more upon the ability to think logically, to handle symbolic and abstract material, and to be capable of lifelong learning." How many of our present programs are geared to producing this kind of people? More than ever before, curricula must stretch the mind, and qualified researchers are already working toward means of making use of more than the present fraction of our mental potential. The Nobel laureate P.W. Bridgman in *Quo Vadis*? pointed out some time ago that "the human race has not yet found how to use its mind," but he felt that we are "on the brink of a major breakthrough." In fact, Bridgman thought that the mark of an educated man was "a realization that the tools of human thinking are not yet understood. . .and that the most important intellectual task for the future is to acquire an understanding of the tools." [14] "Know thyself" may be again the key to modern education in the humanities, but with a different twist. Curiously, one of the aids to understanding how the brain works may come from the monster that many fear will dehumanize us: the computer. Computer research at the moment indicates that we can learn a great deal about the thinking process from studying the way the Machine operates over a period of time and self-adjustment. What a beautiful irony this may turn out to be! There is a moral here too. We are being asked not merely to learn more, but to learn how to learn more. Pre-knowledge training must be a joint operation of humanists and technologists; convincing them and their bosses may take a lot of doing. Louis Armand calls this "a great new discipline of 'scientific and technological humanities. . . which will bring out the qualities of man, the evolution of his mind, and the potentialities of his future." He adds, "It is high time that technology and culture make common cause, and that their disciples came together in one body."[15]

If, as has been said, "each institution shapes its own approach to what kinds of persons it wants to produce," most of the homework and the operational details of new humanities programs will have to be worked out locally. But since we are talking about *criteria,* we may proceed with general suggestions. One might begin with a meeting of representatives from all campus departments to consider such topics as pure and applied humanities, the relevance of the particular vision of the artist and musician to the design problems of the engineer, and the relation of value systems to systems engineering, for example. Professor Allen Rosenstein of U.C.L.A. and others have recently reminded us in the press of the importance of the engineer's awareness of the value systems in the particular environment in which he may be called upon to work, but, more significantly, of the effects of his own value system on that very environment. Sociologists and psychologists could rally round here and furnish tools; historians and anthropologists could remind the general faculty of the truth of Santayana's remark about those who are ignorant of the mistakes of the past being doomed to repeat them—and thus help to build fiber and sinew in the students.

If the humanists and social scientists are to pitch in as they must when a first-rate curriculum is to be set up in science or engineering, planners may have to resort to cajoling them, frightening them, or appealing to their sense of duty. If they are told that many laymen think them the only logical brake to prevent technological forces from racing out of control, making us slaves to an environment we have altered perhaps irreversibly (a common notion in the journals these days), they may well rise to the challenge. If the liberal arts people fail to "apply," the engineers and scientists will try to do the applying themselves. Already, for example, in a course in macro systems at U.C.L.A. there are lectures by Professor Tallman with such topics as "The Human Side of Systems," "Urban Macro Systems" (including the study of waste and pollution), and "Global Socio-economics."

Some of the responsibility for making activists out of aloof humanists will have to be taken over by the graduate schools, which for years have been in a rut of tradition, still in the shadow of the old German doctorate, still requiring Old English and turning out specialists in the works of forgotten authors, with no thought of preparing their students for the actual experience of the classroom and the world, in both of which arenas most of them will spend most of their lives. If they do not take over, they will be lost in the rush, and their men and women will live in a new monastic habitat in a time when involvement is a key word. The eminent geophysicist J. Tuzo Wilson has a word of warning for such individuals: "The only proper person to concern himself with the problems of society is the humanist, but he must learn to concern himself with society in a scientific age and not with the past." Wilson suggests that universities form new departments to train "not scientists but scientifically literate humanists," and that the solution is "to base such studies on a thorough reading of the history of science starting from the earliest times."[16] A close reading of the news, of two or three articles, of even one book like Ralph Lapp's *The New Priesthood* should bring the humanities people off their aloof perch to remind the scientists and engineers of their proper roles and to assume roles of their own for the good of the world. Mere standoff criticism by humanists of technology running wild is not enough; it is like picketing only. What is needed is awareness of responsibility and ability, of sharing, of working in a symbiotic relationship. We have already seen some liberal arts people trying to humanize engineers; some engineering schools are offering engineering for non-engineers. Why stop there? For example, Harvey Mudd is working with another Claremont College, Scripps, an institution for girls, in a joint offering in history of technology.

The engineer is lost today without a sense of involvement in and responsibility for many world conditions; the humanist and social scientist must help him to grow up to consideration of so many matters which, in a happier day, were none of his business. Meanwhile the liberal arts student owes it to himself to take more science, as Wilson suggests. Within the last decade one survey showed that nearly 40 percent of United States college students graduated without a single course in physical science.[17] Today, progress is everybody's business, for good or ill. Science is everybody's business. But progress has its own tyranny. As Robert Gomer has shown in a recent article, man today has the feeling he can accomplish anything, but feels he is being personally left behind, alienated. The price of material well-being is loss of privacy, ugly environment, apathy, a "blunting of sensibilities." Gomer writes,

> Many effects of technological change will be almost impossible to reverse once they have occurred. . . To a much larger degree than we may realize or acknowledge, we are caught up in an evolutionary stream of our making but beyond our control. Even under the best of circumstances we will have to accept the fact that man must change to meet changes he has himself set off.[18]

"Man must change to meet changes he has set off." How else can he accomplish this but through education at the hands of people who know how to discuss privacy, ecology, urban renewal, human despair, and all the problems suggested above: namely, writers, sociologists, psychologists, philosophers, engineers, and technologists working in a unified program?

Students in engineering and in liberal arts must both know the history of technology, its glories, and its present dilemmas: waste, pollution, transport, noise, automation, leisure set up by "progress," population, food, water, and on and on. Both must be informed of the danger inherent in failure to consider all results in any grandiose scientific undertaking which starts out well but gets out of hand, e.g., Project Starfish, as reported by Barry Commoner in his recent book, *Science and Survival.* (A presumably harmless space shot with local and temporary effects turned out to damage satellites in the sky, parts of the Van Allen belt, and radio signals, with effects over a large area and apparently for thirty years to come.) There are too many cases of people doing something because it can be done without asking whether it ought to be done. The fallacy of "If we can do it, we ought to do it, (or we will do it)" can be seen if we recall that technically we can destroy the world. Humanists, biologists, and anthropologists have a duty to speak out here, to include their findings in courses for the scientists and engineers of tomorrow, and to persuade administrators to approve the updating of their offerings. The hearings of the Congressional subcommittee chaired by Representative Daddario of Connecticut are a good immediate step in the direction of benevolent controls; legislation is contemplated to set up some kind of technology review board to think through projects which appear scientifically and technologically sound, but which may have dangerous unanticipated side effects. This is a touchy business, of course; scientists and laymen alike don't like the idea of too much governmental control. At the moment, however, science and engineering, whether they like it or not, are hooked up with government one way or another for as long as anyone can see ahead. Any school of engineering which is considering its program in humanities and in both biological and social sciences will simply not be preparing its students for the modern facts of life if it ignores the implications and cross-relationships of know-how and responsibility in setting up its interdisciplinary courses. Almost anything an engineer does today on any scale has sociological and humanitarian overtones, and he'd better be trained to hear them.

In learning the history of technology and the modern applications of engineering, the student must face the fact that technology is not all good, learn to meet criticism, and learn to separate reason from hysteria in the popular prints. Dr. Emmanuel Mesthene indicated as much when he said that *all* of us should recognize that while technology is liberating, many fear it is enslaving. In an address at the dedication of a new engineering building at Harvard in

May 1966, Dr. Mesthene, Director of the Harvard Program on Technology and Society, pointed out that layman and engineer alike should know as part of their conduct and their human relations that many fear technology destroys some values; that communication and transport made possible by it make have-nots more aware of their state and hence unhappy; that it *can* lead to evil, destroying or poisoning the world; and that it complicates the world and makes people uncomfortable. This was not intended as an appeal to stop technology, but to understand it better, to use it wisely. Ideas like these should be discussed in college classes. Specialist lecturers would have much to contribute. The relations of engineering and philosophy, religion, history, art, literature, government, psychology, etc. can be probed via specialist lectures, small section discussions, and reading in such people as Harrison Brown, Conant, Bronowski, Price, and Mumford. (See Bibliography).

The responsibility of humanities and social science faculties in this age of technology includes self-retraining in the record and problems of engineering and science as part of their own continuing education. As Thomas Huxley and C.P. Snow have told us in effect, we can go on enjoying Shakespeare but we are not educated for our times unless we also know the Second Law of Thermodynamics—speaking symbolically or in principles, perhaps, but none the less truly. Humanist and engineer alike are members of society off the job and should act many roles: personal, professional, societal. The former as a member of society cannot avoid a present choice, either to bemoan the dangers of technology and withdraw to some sanctuary, or jump in and help and even take the lead. In the words of Professor Murray Gell-Mann of Cal. Tech.,

> Society must give new direction to technology, diverting it from applications that yield higher productive efficiency and into areas that yield greater human satisfaction.[19]

Robert Heilbroner goes even further:

> . . .the coming generation will be the last generation to seize control over technology before technology has irreversibly seized control over it. A generation is not much time, but it is *some* time. . .[20]

If Heilbroner is right, the coming generation had better have the best, most realistic, best balanced education possible. Even if he is wrong, or you don't agree with Dr. Gell-Mann, or you would like to add "only" after his "yield," these commentators are no amateurs setting up straw men to be knocked over. These ideas are samples of the yeasty ideas in the air nowadays. How many of them are in the typical classroom in the typical engineering school?

If we shoot for an ideal and content ourselves with the best improvement possible under local circumstances, even if we do fall short a bit, our labors will not have been in vain. A future curriculum in engineering schools or institutes of technology or colleges of science and engineering should have from one-fourth to one-third of its credit hours devoted to humanities and social sciences, in either a four- or a five-year program; it can get along with one-fifth if it must. To be truly adventuresome, administrators might think of forgetting percentages, departments, and schools with their separate autonomies, and talk of *blending* technology and "hum-soc", as it is often called; of thinking through the "mix" of technological know-how and hum-soc monitoring and supply. This effort might involve direct contribution by the engineering or science wings to campus-wide general education setups; the interplay between biology and engineering as in future manipulation of the genetic code, for example (note recent headline: "Biologist Fears Disaster in Efforts to Manipulate Genes"); along with discussions of the phenomena of urbanization (including such social problems as moving Italian-Americans accustomed to over-the-fence family relations in the substandard areas to high-rise apartments where they are clean, efficient, and miserable) and going on to exploring the role of technology in business organizations. It is worth noting at this point that Stevens Institute of Technology, for example, is introducing a degree in social technology. This is in line with a recent statement of J. Herbert Holloman to the effect that the most important aspect of an engineer's education is to become aware of the character and needs of the society in which he will live.[21] If he does become incidentally an enlightened citizen, the engineer will be needed to warn politicians against holding naïve or blind attitudes toward science and technology. We don't want a drift toward the kind of world imagined by Nigel Calder, editor of *New Scientist,* a world in which an honest politician promises in a campaign speech that his audience will have more people, noise, auto accidents, new diseases, supersonic bangs, foreign wars, struggles over ocean riches—a man-made world in which one cannot tell man from Machine.[22]

If readers of this chapter want a good place to begin serious further inquiry, they can do no better than to look up an item called "Education for Innovation," by Daniel V. De Simone of the U.S. Department of Commerce, which appeared in the I.E.E.E. *Spectrum* for January 1968, pages 83-89. The subhead of De Simone's essay gives a clue:

> Practical, creative engineers are desperately needed to solve the complex problems of modern society; yet there is a proliferation of walking formula indexes issuing from our colleges. Perhaps we are educating innovation into oblivion.

Now it is not the purpose of this present discussion of humanities programs in engineering to do much with the technical part of the curriculum. This is well-handled elsewhere by experts. But since I am talking about interplay and relevance in these pages, it is permissible to summarize other points in the excellent work by Mr. De Simone. He points out that invention and innovation are the business of engineering; he says that all consequences of innovation are not good; he would agree with Professor Gomer that it is "unlikely that any society will ever deliberately arrest its technological advance"—hence engineering education had better watch quality of change, but he really comes to the crux of the matter for us when he writes,

> It is an error, of course, to hold that the disagreeable aspects of society are due to science and technology. The fault lies in the manipulators of change, in an attitude of mind *that sacrifices human values for other objectives. Engineering students should know this, for they will be the future instruments of technological change.* (italics mine)

Human values are at the heart of the curricula we have been describing. To ignore them is to suggest that it might be better to drop any serious talk of humanities in engineering. (De Simone, by the way, offers a seven-point program in creative engineering which would fit in very well with a first-rate set of humanities or liberal arts offerings.)

In review, then, and in no particular order of importance, criteria for setting up a viable humanities and social sciences stem in engineering education would include the following ideas for discussion, adaptation, and possible adoption:

1. Knowledge of the history of the engineering profession.
2. Inclusion of hum-soc as a strong part of the major program, not merely as a service area.
3. Interdisciplinary cooperation to the point of blending.
4. Inclusion of traditional courses for personal enjoyment.
5. Inclusion of applied humanities for the good of society.
6. Special attention to creativity, decision-making, and the interplay of value systems.
7. Bringing pressure to bear on departments in the hum-soc area in graduate schools to include some generalist training.
8. Capitalizing on alumni, industry, and government feedback.
9. Heeding suggestions from student-faculty committees.
10. Driving toward concepts of continuing education after B.S. or B.E.
11. Consideration of utilitarian, cultural, developmental, and contextual approaches to a curriculum.
12. Developing capacity in the program to change with change; obsolescence comes fast these days.

13. Campaigning for faculty loyalty and support in engineering.
14. Enlisting more than sporadic help from liberal arts faculty.
15. Defining the local goals and type of engineer to be produced.
16. Making the student see his modern role in proper perspective, providing motivation.
17. Emphasizing design as a meeting place for engineer and artist.
18. Setting up projects appealing to talent, imagination, and creativity.
19. Pounding the engineer's relationship with the needs of society.
20. Accenting flexibility and innovation, but never at the price of loss of human values.

No chapter can do more than suggest materials for a full treatise. And even a treatise on criteria would still merely open the way for optimizing the findings and putting them to work. If criteria are even partly approved, and genuine interest in change exists, organization of detail will follow without too much trouble for specialists indoctrinated in the expertise of structuring. Chekhov is reported to have said that it is enough to state a problem clearly, to criticize without offering the details of solution, since a different type of mind will ordinarily come along and, knowing what the options include, produce a set of answers. In this connection, the specific guidelines set up most recently by the ASEE Report "Liberal Learning for the Engineer" are as follows:

1. Understanding of contemporary change in terms of the role of technology.
2. Acquiring perspective on human life through the record of the past in the humanities and fine arts.
3. Understanding of the contributions of social scientists.

4. Appreciation of the need for continuing education, keeping up to date.
5. Learning how to work with others.
6. Developing personal flexibility and tolerance.

The Report describes the function of humanities in engineering education quite neatly and forthrightly:

> What the humanities provide is what the poet Ransom once called 'the world's body.' They flesh out experience instead of skeletonizing and abstracting. Their concern is with experience, with complexity and ambiguity. They deal ultimately, and in complex ways, with values, with what it 'means' to be a human being. And they sometimes strike the engineering student as remote or merely decorative. Their relevance, however, is in their apparent *ir*relevance. Their 'function'. . .is developmental. They do not seek, however, to indoctrinate; they complicate rather than simplify. As courses and subject matters, they give the student a sense of human life, a sense which they invite him to renew through continuing reflection and continuing contact with the arts.[23]

Off the academic stage and on, one constantly runs into people interested in the problem of humanistic education of scientists and engineers in the present age. A recent symposium in *Technology and Culture* offers a case in point; called *Technology for Man*, it opened with a paper by W.E. Howland on "Engineering Education for Social Leadership," which was followed by three commentaries. Howland reminds us that the late Norbert Wiener years ago stressed the fact that engineers "are also human beings, and as such they. . .*as men* and the society of men which they serve are the ultimate concern."[24] It is furthermore increasingly clear, the author adds, that it is "the engineer's responsibility to see that *technology is for man, and not man for technology.*" In the course of his paper, Howland cites evidence of apathy on the part of scientists and engineers who see problems of pollution, transportation, and despoiling of natural beauty without doing anything about it; he quotes J. Herbert Holloman, president of the University of Oklahoma, on the dangers of a complex society disintegrating if people, especially engineers, are not well-trained in humanities and social sciences, and he brings up again the notion that a good program solution for engineering students might well serve as a model for "bridges over the gulf that divides the continent of the intellect. . .the so-called gulf between the 'two cultures.' " One of the commentaries, by Samuel C. Florman, takes an unusual and provocative view, so fresh that it is like a slap in the face. Maybe this is what we all need. Says Florman, "I submit that study of the liberal arts will rob the engineer of his innocence, destroy his naïveté, stain his character, make him *less* moral. *And this is exactly what the engineer needs, and what will best serve to save our civilization.*" This apparent cynicism is anything but, for the commentator follows up with a disclaimer that we don't want technically brilliant Boy Scouts working from a liberal arts manual. What we do need desperately "are engineers who are intelligent, imaginative, mature, wise, sensitive, articulate, and esteemed, engineers who are 'with it'—in other words, engineers who are liberally educated. Morality, let us pray, will take care of itself." Now all the department heads and administrators have to do is take each adjective in the foregoing list of qualities and figure out what course will take care of providing it.

There is a growing awareness of "what we do need" on both sides of the Atlantic ten years after Snow's first look at the two cultures. In a Dartmouth-inspired conference in Washington, D.C. in March 1967, a panel addressed itself to the overall question, "What can we do to develop a competence and an understanding of science and technology in the liberally educated man?" Out of the ensuing discussion came a modern manifesto: "There must be an appreciation of the relevance of science to engineering with a firm commitment by engineering to the needs of society. A liberally educated man is one who is at home with other men's experience." Looking to the future (and here I anticipate a bit the "thrust" of the next chapter), the conference suggested a unity theme when it urged the engineer and scientist to assist non-engineers and non-scientists to understand the implications of technological change:

> Most of the problems brought about by rapid technological change are social problems, and if they are to be dealt with effectively, require the active collaboration of the scientist, the engineer, the social scientist, and the humanist.[25]

Across the ocean in France, the home of Sadi Carnot and the Second Law of Thermodynamics, the great student strike of 1968 helped to draw attention to the need for reform in education. On September 14, Edgar Faure, Minister for Education, announced sweeping changes which were reported in the Associated Press; the following paragraph is relevant to the present discussion:

> Some 1500 literature students at the Sorbonne will take scientific courses at a suburban university center in an effort to encourage the bridging of academic disciplines. At the same time, science students will be encouraged to take arts and letters studies. This, too, is an experimental program.

And in England something like an academic revolution has been taking place since Snow's *Second Look*, as was mentioned in Chapter I. In the summer of 1968, I visited eight British technological universities at the suggestion of Sir Eric Ashby, Vice-Chancellor of Cambridge University and author of *Technology and the Academics*, a book which should be read by anyone interested in the educational problems we have been examining here. I also had a brief audience with Lord Snow and many discussions with British educators, in the course of which I learned about curricular developments in other institutions which I did not have the opportunity to visit. After returning to the States, I turned in an informal report to my home base, Harvey Mudd College, which had sponsored my trip. Part of that report reads as follows:

> One quickly discovers evidence of ancient aloofness between the cultures as far as formal curricula are concerned in Britain, and much of Snow's early fear can still be documented. This was made plain to me at Manchester and Strathclyde, and I was told this was true at Imperial College, London; as a matter of fact, these three top schools are still called the Three Sisters, although the phrase is dropping out of use. Strathclyde has had four, three, two, and one-year required programs in humanities/social sciences *preceding* elevation

to university status and indeed necessary for such elevation; no humanities/social science is now required of science majors, though some courses are taken as electives. The vice-chancellor of one of the 'Sisters' is reportedly dead set against humanities/social sciences as interfering with scientific concentration and advance; a department head at another felt the same way.

At the other institutions visited, however, there is a different, post-Snow impression. At least ten per cent of the formal program at these schools is either planned or operative in humanities/social sciences; at Bath, the aim is for twenty per cent as it is in the United States. Five schools visited are among the original ten 'new' universities as of 1965 (now shrunk to eight as a result of regrouping), and as such are evidence of a new trend in England counter to conditions when Snow wrote; however, they work with the milk after the big boys have skimmed off the cream of the students. Nevertheless, morale is high, the breach of custom has been made, and everyone feels quality is improving. Many of these schools practice college-industry relations with a vengeance, offering five-year programs with so-called 'sandwich courses'—six months on campus and six off in industry; so far, the feedback from industry favors inclusion of non-technical courses, especially in social sciences. Generally speaking, England's problems in these new technological schools as far as humanities/social sciences are concerned are essentially ours, but different in degree: money, percentage of offerings, concentration versus skimming many subjects, requireds versus electives, faculty-student relations, etc. They are well-acquainted with what is going on over here, envy us our foundation backing, applaud our experiments. City University in London is trying to get a large American corporation interested in support, but this kind of thing comes slowly over there and is not yet in the tradition. Faculty I met know more about h/s in engineering over here than Snow showed, even in his sequel, *A Second Look*. I would favor more interchange between our ASEE and those societies for liberal education in technological institutions which are ardently stumbling along in Britain. The upshot of the research in Britain this summer is that there are definite signs of a change in the intellectual climate of academe in Snow's bailiwick since he wrote (and he would be the first to be happy about this).[26]

And, I might add, in Australia, where Queensland University is studying humanities programs in engineering. Space permits mention of only two interesting programs of special merit in Britain. One is at Loughborough University of Technology, whose 1968-69 prospectus announces the following (p. 57):

> So many University Honours courses in this country have in the past concentrated either on Arts or on Science and Technology that the student was given the feeling of being initiated into one of two quite separate cultures. In an age when our lives are totally involved with technology many young men and women feel that complete specialization on one fragment of either side would unduly cramp the development of their interests and abilities.

The prospectus goes on to describe a four-year course in social sciences and technology which also includes humanities. The Loughborough graduate will hopefully be "the non-specialist who knows what the specialist is talking about," who has "broad education," knows "the importance of critical habits of mind," and "will, in short, be linked with both cultures."

The Bath Technological University brochure, edited by Gerald Walters, gives an excellent account of the rise of the English technological university and the philosophy of unified culture:

> It is, however, with a new university of technology that the real opportunity of finding a coherent relationship between the traditional and the contemporary, between the humanistic and the scientific, lies, since it begins with the fundamental acceptance of the scientific, both in its pure and applied aspects, as the normative activity of contemporary society and with a recognition of the intellectual and social implications of technology. It provides a common frame of reference within which the older traditional activities can find contemporary significance, and science itself recover its role as one of the humanities.

Walters says that the problem of communication between the two cultures can be solved in a generation by changing curricula. The real problem, quoting J.H. Plumb, is that the humanities must change their image, "adapt themselves to the needs of a society dominated by science and technology, or retreat into social triviality."

Only those acquainted with the older class system in England, the nineteenth-century tradition, and the situation as Snow described it in 1959 can fully appreciate the full significance of the new movement in democratic education in these new technological universities. It offers real hope for the future and deserves moral and financial support from all truly concerned with the problems of our technological age.

With this summary of recent developments in educational circles, we are about ready to turn to the wider view, which involves all individuals, especially laymen, who are trying to keep up with technological change and simultaneously lead a full life. Even here, the most advanced universities are taking a hand, as another Snow, John Alan, informs us:

> The dynamics of the science-technology-society relationship is itself a legitimate new field for scholarly inquiry and teaching. . . A common feature of these observations is that university faculties must undertake a major new task of education—of each other! The conditions of the modern world demand a new kind of teacher, as willing to learn as to teach, less parochial in interests and less professional in orientation.[27]

"The key to reform is relevance," J.A. Snow continues—borrowing the word made prominent by recent student revolts, to which the author gives full credit—and "it is time to come to grips with an age." To all of which I add a hearty "Amen!"

We have seen in Chapter II a typical confrontation of two cultures and, in this Chapter III, one field in which points of view may be drawn more closely together, the field of liberal education of scientists, engineers, and humanists. "Changes in education are not going to produce miracles," Lord Snow warns us in the conclusion to his *Second Look,* but if we are lucky, he continues, we can see to it that many of our better minds not stay ignorant of imaginative experience in science and the arts or of their responsibilities to remedy human suffering. But what of those outside the school walls? How can people of goodwill off-campus grow closer intellectually? What are some possible steps toward a more unified, richer culture? Why should anyone, for that matter, bother to make the effort required to think or appraise personal goals when it is easier to turn on the TV or risk the boredom of an almost purely materialistic existence? I shall try in the next and final chapter to show why the gulf between the cultures should be bridged before it is too late, what kinds of bridges may be constructed, and what the bridge-crossers may find on both sides of the gulf to make the trip rewarding.

Chapter IV

Toward One Culture: A Wider View

"Closing the gap between our cultures is a necessity in the most abstract intellectual sense, as well as in the most practical."
–C.P. Snow

Turning from one formal, structured program in engineering education as an example of gap-bridging to a general or self-educating program, I should like to argue that even in this age of specialization and technological stewardship it ought to be possible for scientists, humanists, and laymen generally to lead fuller lives and to do it together, not apart. Why indeed should so many people feel that this is an either-or world, that one has either a scientific mind or an artistic "bent," or a literary disposition, and so on? May one not lean in one direction most of the time and yet tilt in another now and then, to restore balance? Literary specialization of the kind still fostered in graduate schools, which offers little in the way of experience that carries over into the classroom, is an anachronistic luxury these days; scientific specialization may lead a man to the top of his profession, in lonely exile where his own colleagues cannot follow, but at what expense? In both cases, literary and scientific, what may have happened to the man? We have asked in the last chapter what kind of engineer we want to produce. Now we may ask what kind of human being we want to live with in the near future.

Once more at the risk of sounding like a meddler or lay evangelist, I would suggest that any specialist has the right to pursue his specialty on the job or in private, but that he also has the obligation, to himself and others, to exist as richly as possible as an off-duty human being. Education for a new humanism is not only relevant but necessary; it can be done by scientists or humanists, but ideally together. The alternative is to run away and hide or to write immature self-pitying novels, as we have seen. Far better to construct a new mythology, I would think, along the lines of Ruth Benedict's observation, "Man in all his mythologies has expressed his discomfort at a mechanistic universe and his pleasure in substituting a world that is humanly motivated and directed."[1] Luigi Barzini, author of *The Italians,* in the course of a TV special, March 11, 1969, on how the city of Florence salvaged her flood-damaged frescoes, made the statement that "today we are great technicians and small believers." To what extent is the latter a product of the former? Now if a scientist or humanist or layman isn't completely sold on the idea that the two cultures are moving away from each other and must remain apart, how can he go about becoming a believer? Is there any evidence or argument for the opposite point of view? And just why should Lord Snow have thought the polarity he observed to be a danger to the world?

I have come to believe, with the help of several scientists, writers, and artists, that there is an equally strong argument for unity and order at the heart of what inevitably seems like a splintered culture on the surface and that, given some interest, imagination, and effort on the part of the intelligent people we have been dealing with throughout this book, some kind of belief, mythology, functional illusion, or reason for being and enjoying being alive can materialize, and on a shared basis. Even today we can preserve what's left of Nature, get closer to it instead of farther and farther away; even today in the U.N. politically and macrocosmically and in local cultural areas individually and microcosmically, we can and must have better understanding of others or face the failure of both worlds; and finally, as individuals we can stop the drift to only material things (without giving them *all* up) and develop a larger fraction of personal potential through instructed awareness of what has been passed by in the rush for the world's goods. The restoration of Renaissance man may be an impossibility, but we can still resist becoming automata and, in an attempt to be closer to the whole man idea, avoid the boredom and alienation so characteristic of our times. It won't be easy. Kenneth Keniston, author of the stimulating work *The Uncommitted,* outlines some of the hopes and pains of such a course:

> 'Fulfillment' depends on individual potential and on social opportunity; human 'wholeness' depends on what there is to be made whole. . . .A whole man or woman has the capacity for zest, exuberance, and passion, though this capacity may often be in abeyance. . .retains the capacity for openness, sensitivity, and responsiveness to the world around him: he can always be surprised because he remains open to that which is alien to himself.[2]

"Open to that which is alien": the first requirement in the breaking down of the selfish two-culture concept. The next is the personal commitment necessary to produce a new society which is closer to one culture, but retains diversity within the whole. Keniston continues,

> There are millions of men and women who sense vaguely that something is amiss in their lives, who search for something more, and yet who cannot find it. Their idealism will not be easily redirected to the creation of better lives in a better society; it will require imagination, vigor, conviction, and strong voices willing to call for many years, before we *dare raise our aspirations beyond vistas of total technology to visions of fuller humanity.* But for the first time in American history. . .it is conceivable that a whole nation might come to take seriously these ancient and honored visions. . .(italics mine)
>
> We can hope for such new commitments in the future only if men now begin to resolve their alienations by committing themselves—through the analysis, synthesis, and reform of their own lives and worlds—to the preparation of such a new society, a society in which whole men and women can play with zest and spontaneity, can work with skill and dedication, can love with passion and care—a society that enjoys diversity and supports human fulfillment.[3]

The possibilities of such a society or culture should be exciting enough for those who rise to challenges. But there must be better communication among people who are intelligent, alerted, articulate, dedicated. They must be shown the intellectual pleasure of recognizing similarities between apparently dissimilar entities, a metaphysical process which will reveal, for example, that the humanist and the scientist are similar in many ways, but not identical, nor should a unitary culture mean identicalness. The alternative to Keniston's whole man has been around as in the mass-think under Hitler; the automatization of people is a reality on the assembly line and threatens to move ever farther afield, as seen in the conformity trends of modern advertising. This alternative is horrible to contemplate. As Floyd Matson tells us, it eclipses the old liberal vision of the whole man and leaves instead a radically broken self-image; this breaking, in turn, "parallels the disintegration of the inner sense of identity, the flight from autonomous conduct to automaton behavior in the modern world."[4] In contrast, Lillian Gish, in the course of an interview in April 1969, described the improvising techniques of the actors in early movies who had only an outline of the plot sequence, no memorized lines to mouth in those "silent" days, and no "method" training. The result, as she put it, was "It was much more fun in those days to work than to play."

Since we have been speaking of awareness and commitment, perhaps this is as good a place as any to begin discussing ways of coming together through others' experiences. Lynn White would go beyond the engineering curriculum described in Chapter III, although conceding some value to reading Plato and contemplating Matisse, and would recommend to all of us, engineers and laymen alike, the study of the history of technology as basic equipment in learning to be "vividly aware of the new atmosphere which all of us. . .are now breathing whether we know it or not." He calls upon engineers to "join vigorously with a small but growing group of professional humanistic scholars to build a bridge, already well under construction, between their traditionally separate activities. That bridge is the history of technology and science."[5] It is White's feeling, to which I heartily subscribe, that the contemporary student of science or technology can best understand his relationship with his environment and his place in it by reading of the involvement of his professional forebears "in the general intellectual, artistic, religious, political, economic, and social milieux of their times."[6] I would go further and recommend this study to the humanist, artist, and layman as a first step in the informal education process.

Louis Armand and Sir Eric Ashby (as we have seen in the two previous chapters) actually see technology itself, not merely the study of its history, as the bridge or cement between the two cultures of science and the humanities. In reply to the question so pertinent to our present consideration of the wider view, "Does not technology in fact form an integral part of humanity in the widest sense of the term?" Armand replies,

> The world of tomorrow will realize that it does. The world will no longer say to those who fail to cross the *pons asinorum* of mathematics at the outset: 'You are arts men, confine yourselves to the universe of letters and from now on say good-bye to the universe of realities.' On the contrary, it will direct their steps towards a great new discipline of 'scientific and technological humanities,' a 10-times-magnified lesson in everyday things with a philosophical angle, which will be accessible to all and which, no less than classical philosophy, letters, or history, will bring out the qualities of man, the evolution of his mind, and the potentialities of his future.[7]

Sir Eric reaches the same conclusion by a humorous example in his *Technology and the Academics.* Developing the point that a man should follow a path to culture through his specialty, not by-passing it, he outlines the possibilities facing a student of the brewer's art:

> So there must first of all be an assurance that the student genuinely wants to make beer. From this it is a natural step to the study of biology, microbiology, and chemistry: all subjects which can be studied not as techniques to be practiced but as ideas to be understood. As his studies gain momentum the student could, by skillful teaching, be made interested in the economics of marketing beer, in public houses, in their design, in architecture; or in the history of beer-drinking from the time of the early Egyptian inscriptions, and so in social history; or, in the unhappy moral effects of drinking too much beer, and so in religion and

> ethics. A student who can weave his technology into the fabric of society can claim to have a liberal education; a student who cannot. . .cannot claim even to be a good technologist.[8]

If skillful teachers like those Sir Eric has in mind are hard to come by, I would suggest that most mature readers could find their way through the ramifications of beer drinking with his outline alone as a guide. It might be a good experiment to convert those who doubt the interplay between the so-called two cultures. History of technology, technological humanism—these are but two methods or labels from a long list to be considered here, but they may serve as a logical transition from curriculum tinkering to informal pursuits open to any curious reader. The goal will be the same in any case; the end result may be reached in many ways, all of which may differ in emphasis or approach, but not in kind or purpose.

Speaking of teachers and pupils, Howard Mumford Jones comes up with two practical suggestions, as he calls them, which might, like Ashby's, prove a bit difficult to put into effect. Nevertheless, both men have a point and they are moving in the right direction. Jones proposes first that we reduce the number of admissions to graduate schools offering training in the humanities in order to "weed out those blameless mediocrities now furnishing too large a fraction of our Ph.D.'s." His second idea, "equally simple, equally revolutionary, and equally logical," is that if we are talking about the whole man, we had better make sure "that we have the whole man before us to educate."[9] Jones' program for admissions officers would call for less attention to the number of A's and more to the wholeness of personality indicated in the records of student applicants; educators, in an effort to cut down the number of unhappy bookworms, should also look for evidence of humor, musical skill, interest in sports, and good taste in food, drink, and manners—thus hopefully avoiding excess specialism and making humanistic studies human. Again, there is a program here for the layman who is working on the definition of general culture.

P.W. Bridgman, mentioned in the previous chapter, is not so concerned about kinds of teachers and courses as he is about learning how to use our minds properly. Calling for courage and vision, he too emphasizes the necessity of scientist or engineer and humanist working together. Bridgman feels that science is beginning to find some answers and that the humanities, because of their involvement with verbal machinery, must pitch in and help:

> I would place as the most important mark of an adequately educated man a realization that the tools of human thinking are not yet understood, and that they impose limitations of which we are not yet fully aware. As a corollary it follows that the most important intellectual task for the future is to acquire an understanding of the tools, and so to modify our outlook and ideals as to take account of their limitations.
>
> This task is not to be accomplished by any 'return' to the insights of the past. The insight that there is any problem here at all is devastatingly new in human history. *The sciences and the humanities find themselves facing the problem together; it is too difficult and too pressing to permit the luxury of a division of forces.* [10] (italics mine)

How many who have read Snow have thought of the two-culture philosophy as a luxury? Perhaps our best approach to understanding is on the basis of what we can afford?

All the writers quoted in this chapter thus far have in one way or another called for the whole man, not separated men, specialists, alienated human beings with broken self-images, bookworms, or intellectual snobs. Lewis Mumford speaks for them all when in "In the Name of Sanity" he tells us,

> Not the Power Man, not the Profit Man, not the Mechanical Man, but the Whole Man. . .must be the central actor in the new drama of civilization. . . .If technics is not to play a wholly destructive part in the future of western civilization we must now ask ourselves for the first time, what sort of society and what kind of man are we seeking to produce. [11]

It should be obvious by now that the whole man cannot exist in a two-culture context. In an age of technology those convinced of the need for Mumford's central actor will have to commit themselves to some kind of personally- or community-oriented activist program. Tempting as it must be to contemplate isolationism, to run away, this is a weak way out, a sheer escape. Beautifully written though it is, Josephine Johnson's recent best-seller *The Inland Island* is just such a lure to escapism, not a facing of uncomfortable fact or a rallying to defeat the enemy. Granville Hicks describes the author in a review as "fortifying herself against the society that has been created by technology run mad" and reminds us of the reactionary romantics who "never doubted that men were more important than machines."[12] The cruel and beautiful Nature Josephine Johnson describes can yet be saved, but hardly by an ascetic philosophy. At the other extreme and also quite recently, Erich Fromm, in *The Revolution of Hope,* agrees that "in the one-sided emphasis on technique and material consumption, man lost touch with himself, with life," but calls for regional and local councils or clubs to work "toward a personal transformation from an alienated person into one of active participation." Ashley Montagu, reviewing Fromm, rightly points out that "His appeal is to those who are concerned enough to be willing to become involved."[13] At any rate, the choices should be clear.

Whether technology is friend or enemy will depend on who handles it or controls it if it is left alone, but such divisiveness can be avoided by union in a common field, as between artists and engineers (topic already touched on by Overbeck and Davis in Chapter I and briefly noted by me in Chapter II). Architects and engineers, as we have seen, meet happily in the field of design. And Overbeck further notes:

> . . .a promising trend catching on internationally among artists and technicians. They are carrying on a direct dialogue now, and exploring potentialities of new art forms in the skills and associative abilities each brings to this meeting of what Snow considered incompatible minds.[14]

Gyorgy Kepes, head of the advanced visual design program at M.I.T., to whom we shall return again for his valuable aid in the important business of acquiring new vision as a basic component in understanding unity in nature, has been campaigning for this kind of dialogue for some time. In "The Visual Arts and the Sciences: A Proposal for Collaboration," a piece which has appeared in at least two magazines and one anthology, Kepes has criticized modern artists for missing "possible vital connections with contemporary intellectual and technological reality" in a plea for one grand culture:

> Only complete acceptance of the world that is developing can make our lives genuinely acceptable. Such acceptance involves two tasks: to advance in every field to the furthest frontiers of knowledge possible today; and to combine and communicate all such knowledge so that we gain the sense of structure, the power to see our world as an interconnected whole.[15]

Art can provide this sense of structure. We may not all, like Blake, become capable of seeing the entire universe in a grain of sand, but we can work at it. Kepes' ultimatum, you will note, allows for the individual within the general, the specialist along with the universalist, the one culture in which apparently (superficially) opposing forces can work in symbiotic relationships.

That art can help men to find an answer to the challenge of science is a point already raised by Rabinowitch in Chapter II, Section 7. For further insights into aesthetics and the remarkable similarities among the creative experiences testified to by outstanding men and women, the reader should go to a volume like Ghiselin's *The Creative Process,* where, for example, he can become acquainted with the dream experience of the mathematician Poincaré, who was overwhelmed by the beauty of a new concept.[16] Bronowski has remarked on the similarities between a poem and a theorem. It is not unusual for a scholar in the higher reaches of mathematics to talk about a beautiful, as opposed to an ugly, equation; in fact, such a man is likely to consider mathematics in the humanities, not a science at all. Likewise in the rarefied atmosphere of decision theory one encounters the notion that a decision between two alternatives may be said to favor the prettier one, in terms of a neatness, symmetry, pleasantness of form which subconsciously sways the brain as it grinds out the answer. Scientists and engineers in seats of power could do worse than investigate these borderline activities, these gray areas where two cultures meet. As a matter of fact, Lionel Trilling, in commenting on the Snow-Leavis controversy, touched on exactly this point:

> The aesthetic mode is integral to the idea of culture, and our judgments of social groups are likely to be made chiefly on an aesthetic basis—we like or do not like what we call their life styles, and even when we judge moralities, the criterion by which we choose between two moralities of, say, equal strictness or equal laxness is likely to be an aesthetic one. [17]

Snow has remarked on the danger inherent in a widening split between the two cultures. John Dewey had sensed this forty years earlier, and he too had stressed the interplay between art and science which I am bringing out at the moment. Here is Dewey's warning summary:

> Surely there is no more significant question before the world than this question of the possibility and method of reconciliation of the attitudes of practical science and contemplative esthetic appreciation. Without the former, man will be the sport and victim of natural forces which he cannot use or control. Without the latter, mankind might become a race of economic monsters, restlessly driving hard bargains with nature and one another, bored with leisure or capable of putting it to use only in ostentatious display and extravagant dissipation. [18]

Written fifty years ago, these words have the ring of prophecy and the odor of contemporaneity at the same time. While Dewey's remarks were still fresh, J. Arthur Thomson bemoaned the misunderstanding which was largely responsible for what he called "the opposition between science and feeling." He felt that modern science had added "thrilling impressions of manifoldness, intricacy, uniformity, inter-relatedness, and evolution," and agreed with an unnamed philosopher that science was truly "one of the humanities." [19] Inter-relatedness? Science one of the humanities? Voices crying in the wilderness apparently. One wonders, with such vision and wisdom available, where things went awry.

Here and there today the common reader can find evidence for these truths, but they need more publicity, and here is where awareness and commitment come in again. Anyone who has read the recent best-seller (later made into a rather good film) *The Sand Pebbles,* by the late Richard McKenna, can hardly forget with what sensitive detail the author showed in the introductory passages the love of the engineer for the Machine he cared for and served in the bowels of the humble river gunboat. Man and Machine were one, the beauty of a well-oiled and rhythmically functioning power plant was obvious. Or one may turn to an avant-garde engineering journal like the English publication called simply *Engineering* and document the newer thinking in terms of unity which is so essential, not only in design, where utility and beauty can and must meet, but in solving the problem of ignorance and two cultures. Here, for example, is an account of a conference of engineers, artists, and designers held in Britain in March 1964. Two conferees ask how one can communicate through a motorway, or a gasworks, a spoon, or a bulldozer:

> In answering this question they come as near to restoring unity to the two cultures in engineering design as anyone has ever come. They suggest, first, that the things we are called upon to design are to be considered to a greater or lesser extent as part of a wider unity—of the town or of the countryside; of street, room, workshop, or valley. [Then follow the ideas mentioned earlier in this volume, that a motorway should enhance, not ruin the landscape; a factory add dignity, not ugliness, etc.] They suggest, secondly, that our products are designed for use, but the designer can employ the devices of art to make the use more apparent and therefore more convenient.[20]

The article continues by saying that "they" conclude designers must understand customers and realize that beauty is more than outer appearance. Granting importance to materials and technique, "they" must come to realize the need to know and understand "man himself, his physiology, his psychology, his reason, his emotion, his conscious and subconscious being." A large order, this, but better to assimilate the Machine than to prostrate oneself before it or hurl oneself on it in despair. "To know and to understand" is a brave phrase. But we have shown enough perhaps to indicate the possibilities of art and design as one of the fields where two opposing parties can meet. Let the curious reader follow up this lead for himself. He might well begin by looking up Lewis Mumford's *Technics and Civilization*, turning to the section called "Assimilation of the Machine," and reading with particular care the subsection titled "The Esthetic Experience of the Machine."[21] Mumford will make him see a skyscraper or a crane in a new light as he traces the Cubists' overcoming of the "association of the ugly and the mechanical" and comes down to artists like Moholy-Nagy and Gabo, who have "created in form the semblance of the mathematical equations and physical formulae that had produced our new environment."[22] New ways of seeing and feeling are required here. These will be the subject of the following section of this discourse, forming as they do the fundamental requirements for travel back and forth across the borders of the two cultures. I can only hope that one day soon no passports or visas will be necessary.

What has been done jointly by art and engineering may serve as a model for other adventures in unity between scientists and humanists. New ways of seeing, feeling, and using words are integral to such undertakings. We may have to replace Auden's home, car, radio, and frigidaire with meditation, walking, looking, and touching. And, again, the point must be stressed that we do not wish to make a scientist into a humanist or vice versa. But if, as some wag has remarked about current theories of light, it is a wave on Mondays, Wednesdays, and Fridays and a particle on Tuesdays, Thursdays, and Saturdays, with Sundays off, we might think of people as humanists part of the time, scientists part of the time, and both part of the time. Slowing down the pace of modern living is very important in learning the ground rules of communication between cultures. Ask a student today how long it has been since he lay on the grass in his back yard and watched the ants, and he may look at you as if you were demented. Suggest that he learn to hear more delicate patterns of sound after being bombarded with noise in a discotheque, and he may say you are not "with it." Try to get a young engineer used to precise tolerances or "correct" form to see one object in more than one way, and you may have a problem on your hands. But if Bridgman is right when he says that we have just begun to know how to use our minds, and if Dewey is right when he says that science and aesthetics need each other, and I believe they are, both humanist and scientist should at least be sporting enough to learn a bit about each other's life styles. This involves retraining vision.

Strange things happen when the layman stares at his first Picasso with its figure looking in profile and straight on simultaneously. Or when he comes upon Moholy-Nagy's *Vision in Motion,* whose photographs force him to *see* in more than one set, obvious way. The imagination that a poet or artist uses characteristically may seem odd at first to a person taught to work with formulae to find *the* answer. I have frequently been amused when students of science or engineering encounter Wallace Stevens' "Thirteen Ways of Looking at a Blackbird." To many of them, as Gertrude Stein might have said, a blackbird is a blackbird is a blackbird. Or a tree is a tree is a tree. I recall a classroom experience in which one engineer had great difficulty in describing a tree as seen through the window. What was there to say, let alone write about? Another student, a math major who later got a double degree in philosophy, began to coach him, and others joined in, "seeing" the tree in various ways. Before we were all through we had touched on the form, color, height, and spread to begin with, and then had branched out, if I may say so, to the details of the tree that would be obvious to a botanist, a woodsman, a lumberman, a student seeking shade, a gardener who had to rake underneath, a bird seeking a landing field, representational and abstract painters, and even an English professor mindful of Marvell's "green thought in a green shade." Some were still unconvinced, but for many even such a simple exercise was an eye-opener, the kind of eye-opener that is the beginning of real communication.

At this stage I would recommend the interested members to go to such a book as Gyorgy Kepes' *The New Landscape (in Science and Art).* [23] This work is a "must" for anyone interested in seeing and in considering the evidence for order and design in nature and their effects on both primitive artifacts and modern commercial reproductions. Photographs of cross-sections of twigs reveal basic patterns of symmetrical design; photomicrographs of the surface of a snail's tongue yield line and structure almost beyond credulity. A first step in such awareness is to realize what one has missed in the apparently obvious and familiar around him. The next, as Kepes shows us here and in the experiments at M.I.T., is to realize the creative potential of a stimulated imagination which can combine nature and science in producing new forms, static or moving, in experiments involving light, color, and motorized parts.

The moral so far is, of course, that while all of us look, some of us do not see, and that when we see, we do not necessarily see the same things or in the same way. It still comes as a shock to some students that a humanities class will come up with many different interpretations of one poem, or an art group on the campus with many different versions of what (photographically) is obviously only one thing that anyone can see, the college library. This is an analogue of social and political experience as well, as anyone will recall from his first trip abroad—if he is lucky enough to have barriers swept away in this manner. Dubos, speaking from the viewpoint of the biologist, is quite right when he says,

> The physicist, the biologist, the humanist, and the layman can all find a common ground for discourse if they talk about matter, life, or man as perceived by the senses, or as apprehended in the form of images, analogies, and responses . . . .Specialists must return to the original human basis of their work if they want to converse with mankind. [24]

The perception of the senses, however, often calls for training or re-training in otherwise intelligent observers who have been in a professional or specialized rut. Tolerance between cultures depends on facts at one's command and attitudes toward those facts. Awareness demands openness of mind plus curiosity. It is not difficult to demonstrate that we see what we can see or want to see or that many of us, if not most, go around a good deal of the time as if we were wearing blinders, like horses.

The story of the college class and the tree outside the window is a parable of the ways of seeing. Or a fable, if you prefer, illustrating the vision difficulties of two cultures which have to learn how each other sees. If perception is adding meaning to sensation, a good deal depends on the stock in the mind of the person doing the looking. Another story, a true one, which illustrates the problem, was recounted years ago in *Reader's Digest* by Donald Culross Peattie. He was discussing the Joshua tree, described by Webster as "a tall branched arborescent yucca." There are spaces in the Southwest, along Route 66 and elsewhere, where utter barrenness seems to prevail except for the scattering of these ungainly but interesting trees, which to those with a comic sense resemble drunken hatracks and to those with a touch of the poet, ghostly sentinels, especially at night. To the average traveler, however, the area is bleak nothingness between inhabited communities. Peattie showed his readers otherwise. He described the work of a certain moth which influences the reproductive process and then the battle between a growing tree and a type of termite which can ingest wood but not digest it. Another creature, living off the termite, in return for board and room, "cracks" the wood for the termite so it can digest. If the tree is strong enough, it will endure in spite of the termite, but now and then the battle goes to the insect, and one or more of the Joshua's ungainly arms will blacken and droop, with results visible even from the highway. Who can now drive by after learning this and say to his companion, "Look at all that emptiness—there's no sign of life except those crazy trees—absolutely nothing going on."? To the *informed* eye and brain there is now a whole drama of life and death going on out there with amazing biological and chemical actors. And if, in the spring wildflower season, the average traveler would turn off the road for a minute, lie on the desert floor, take out a pocket lens and study the almost microscopic beauty of a plant an eighth of an inch high, he would see what might have been missed right before his eyes.

To see the way another man sees is important in bringing people of one culture together with those of another. If there are no Joshua trees handy, reading, conversation, visiting art galleries, listening to music can more conveniently take their place while they fade to the status of symbol. Along with seeing, while we talk of senses, should go, of course, ways of feeling. And here, scientists, engineers, and technologists sometimes find it difficult to set aside the rational and the measurable because these things are basic in their work. Non-verifiable truth, play, emotion, they feel, should be left to swamis, children, or poets. This is an unfortunate state of affairs, and one book can hardly alter it much; however, I should like to send them (and our curious layman) to S. Giedion's *Space, Time and Architecture* for preliminary instruction and inspiration.[25] Among the many ideas in this large and profusely illustrated classic are the following:

1. Even to ask what science and art have in common indicates the lengths to which specialization has gone, and the unfortunate separation of thinking from feeling.
2. In recognizing likenesses and differences among fields, we must return to a general pattern of culture.
3. Industry, developing, unconsciously creates new powers of expression; the architect and the engineer should never have been separated (nineteenth century), for the Machine has waked up the former and the constructions of the latter have demonstrated that they are artists.

Giedion returns to our ideas on seeing by tracing the historical changes in attitudes toward the Eiffel Tower as the public learned more about vision. Then he turns to feeling. An integrated culture, he says, produces unity of feeling among technical and artistic people, but a disintegrated culture divorces them, killing the normal interplay of emotion, leaving only the functional. We must find harmonies between inner states and outer surroundings by opening new realms of feeling. "It is a long way from the specialization of today to the reinstatement of a universal viewpoint, in the absence of which any true culture is unthinkable. It lies in the incorporation into science of the realm of emotion as it is expressed in art."[26] Giedion restates some of the points raised here: We must learn more about the creative process in various fields; feeling affects decision; the individual must be an integrated person, not just a specialist. His main contribution, however, unabashed and firm, is to insist that feeling today is more difficult than thinking and that we must set about training it. The outstanding task of our period, according to this eminent architect, (and it must be obvious now why I have borrowed so much from him) is to humanize—that is, to reabsorb emotionally—what has been created by the spirit; emotional and intellectual outlets must not be separated. Giedion too is on the side of the affirmers. He very definitely feels that our period can cure its disease by recreating the whole man—at a price. What I have referred to as interest plus work or commitment plus involvement comes out in *Space, Time and Architecture* as an awakening cultural consciousness requiring a will for inner change plus forward-looking preparation. If the reader adds up Kepes and Giedion, speaking for many others as well, the sum for our times is expressible in the motto, "See better and feel more."

In his important study of alienation in modern life referred to above, Keniston sees the underplaying of feeling as a large contributing factor to maladjustment. He finds that the "preferred techniques of technology involve two related principles: that we *give priority to cognition,* and that we *subordinate feeling.*"[27] As a result,

> . . .feeling as a force of independent value—all of the passions, impulses, needs, drives, and idealisms which in some societies are the central rationales of existence—are increasingly minimized, suppressed, harnessed, controlled and dominated by the more cognitive parts of the psyche.[28]

It is not surprising to find the insistence on analysis of problems relegating discussions of why and what "to neurotic adolescents and to a few artists and dissidents." And since analysis "almost inevitably involves reduction" in an attempt "to divide unities into fractions," it is further not surprising to find that "principles which might help explain wholeness, intactness, integration, and unity are rarely articulated."[29] Keniston continues,

> The dreamer, the visionary and the poet are thought to be of little use in the day-to-day working of technological society; personal expressiveness and 'style' are of secondary value except as entertainment. . . .other areas of life, family and leisure, are almost invariably relegated to a secondary role, termed outlets, recreation, havens, or exceptions to the basic rules of our social order. . .–they are the froth, frills, safety valves, and status-symbols a technological society must allow itself.[30]

Thus, indirectly, Keniston emphasizes the need for feeling which Giedion proposed directly. But in his association of loss of feeling with growing alienation in our day, the former is more frightening. If the technologist will accept Keniston's findings as proper information, it lies easily within his power to do two things simultaneously by adjusting priorities: through new awareness and a sense of responsibility to restore balance to modern life and reunite two ways of looking at life—two cultures—so that both are enriched. The message is clear. The responsibility is indicated. Time is running out. Charles De Carlo, quoted in Chapter I, nevertheless remains sanguine: "The idea that rationality is the end-all of life is giving way to a more humanistic approach, more concerned with ethics and quality of life."[31] Intuitively I must agree and hope that he is right—and hope that more evidence accumulates, fast.

As a corollary to absence, loss, or downplaying of feeling, accompanied as it is by an exaggeration of the importance of the rational, Erich Fromm has recently noted a presence of emotion wrongly directed or involved:

> The tendency to install technical progress as the highest value is linked up not only with our overemphasis on intellect but, most importantly, with a deep emotional attraction to the mechanical, to all that is not alive, to all that is man-made. . . .Those who are attracted to the non-alive are

> the people who prefer 'law and order' to living structure, bureaucratic to spontaneous methods, gadgets to living beings, repetition to originality, neatness to exuberance, hoarding to spending. They want to control life because they are afraid of its uncontrollable spontaneity. . . .[32]

In order to come closer together, then, both cultures must redefine the familiar in terms of seeing and feeling. As Lewis Mumford put it in 1948, in discussing the repressed functions of the modern personality,

> This means that we must reverse the order of development which first produced the machine: we must now explore the world of history, culture, organic life, human development, as we once explored the nonliving world of nature. We must understand the organics and psychics of personality as we first understood the statics and mechanics of physical processes.[33]

Probably both sides know this, but like political bodies, no one wants to give up any territory. Again, as in politics, they had better sit down at the arbitration table while they still can. And, as in labor relations with management, a skilled mediator is necessary. As Peter Viereck, poet and college professor, sums it up so neatly,

> To mediate between the two worlds of machine and spirit is, to be sure, not the same thing as unifying them. It is merely a gesture of unity, a vivid metaphor of the solution, not the solution itself. It will solve neither the moral problem nor the economic problem–not even a paranoid Muse with delusions of grandeur can claim to do so–but it may inspire in society that bifocal vision which is the prerequisite to any solution.[34]

The need for unity calls for some evidence of it, to convince the skeptical that bifocal vision is worth the trouble. Among scientists, Lord Brain is a fine example of one who has found a double life rewarding; he may be expected to write on the nervous system or Shakespeare in almost any issue of *Lancet*. In *Science and Man* he suggests that the social scientists may provide the bridge between society and scientists, thus adding another possible approach to our growing list. As far as I am concerned, there is more than one answer, and any one that works, any three or four that work, will be welcome. The point is that there are bridges, and people are waking up to the fact. For one thing, Lord Brain, quoting Stubbs, shows the humility required in a meeting of cultures:

> It is arguable. . .that with man's invention of the H-bomb and his growing influence over his own environment, intelligence has become a factor threatening, rather than enhancing his chances of survival.[35]

He endorses what Commoner has made almost a platitude, the need to foresee the consequences of new work in technology. He sees another need for a unifying world view of the nature of man, one culture, in which persons are regarded as values in themselves, and the public knowledge of science can be integrated with the private experiences of people:

> I suggest that one of the essential ingredients of such a view should be the primacy of the private, personal, subjective, individual experience over any public account which science can give. This means that persons are to be regarded as values in themselves, and not as reducible to either physicochemical systems or bundles of psychological trends or impulses. The social and political implications of this are important. One of them is that science, though an end in itself to the scientist, is only a means to an end where other people are concerned, that end being the possibility of their greater fulfillment as persons.[36]

We are back to ends versus means again. But what a beautiful, heretical, balanced, humane view!

Other top men are coming around to a similar view. As Brain is careful to distinguish ends from means, Glenn T. Seaborg, Chairman of the U.S. Atomic Energy Commission, reminds us of the difference between knowledge, which specialization gives us, and wisdom beyond knowledge, for which the world cries out today. In a speech later revised as an article, Seaborg writes about the uncertainty of modern life in a manner we have already found to be characteristic of artists and playwrights: he senses

> an increasing uneasiness about the state of our personal and community lives in a highly materialistic society, a concern

> over the individual's role in the growing complexity and impersonalization of that society, a groping for 'national purpose,' and a feeling that the unity of man, referred to by poets and philosophers throughout the ages, is becoming a reality with immense and physical implications.[37]

Brain and Seaborg have the wider view which senses order within chaos, an order apparent to an artist like Kepes and, in evolutionary perspective, to a biologist like Dubos. They are scientists not blinded by their specialties. Another is Lancelot Whyte, physicist and lecturer, British expert on jet engines, who has written of a "tendency toward form" in the universe observable in the inorganic world "in the genesis of spiral nebulae, of our solar system, of crystals, of molecules," and in the organic world, including the "ordering processes of the human brain-mind."[38] Whyte uses phrases like "end of the age of separation," "tendency toward order," and "the unity and continuity that run through life." He calls his explorations philosophical and *pre*-scientific, crediting architects with being the first professionals to spot this tendency. Against any possible reader objection that we are flying off into mysticism, I quote Whyte on this last point. Can anything be more straightforward than this?

> Is the architect an imaginative creator of beautiful buildings, a calculating engineer of structures, or a planner of homes enhancing family life? Some claim he is all this and more: a humanist unifying all aspects of life in a manner suited to, say, ten or twenty years hence! Thus they welcome a philosopher of form who tries to make his thinking integral with science and is looking ahead.[39]

It is not surprising to find Whyte, in this article which must be added to our basic reading list, decrying the emphasis on reason alone and calling for more feeling to put reason in its proper place and thus strengthen it. He would agree with Keniston on the need in man for emotional excitement, arguing that reason cannot imagine, judge, or select; imagining springs "from a deeper, less conscious level than that of intellectual processes and rational analysis. . . .everything that we value springs from partly unconscious formative or ordering processes in our minds." This combination of intuitive and animal vigor ought to please artists and writers. It ought to excite anybody. It reminds us, and let the scientists and engineers heed this scientist, that man is, in Dubos' phrase, "so human an animal," and should remain so.

As Whyte moves on from the humanist concept of an architect as artist, engineer, and sociologist all at once to the ultimate greater unity of global community, I am led to make a transitional leap to the general notion of a growing recognition of the unity of knowledge. As Karl Jaspers sees it, the very existence of a university "stands for that wholeness and oneness of all knowledge" enabling one to know in the broadest sense of that term. Jaspers defines a university as "simultaneously a professional school, a cultural centre, and a research institute." How, he asks, "is it possible to divorce, say, design in engineering from the philosophical consideration of aesthetics, or to study architecture away from sociology or psychology? How can applied biologists working to resolve problems of world hunger ignore ethical and religious attitudes to population control?"[40] In reply, I must say that I do not know how it is possible to divorce or ignore, but that it happens because too many people still are handicapped by the blinders of education outmoded and have not grown mature enough for bifocals. A philosopher like Jaspers can see the essential tie-ups, but until he gets into more reading lists, the social lag will continue, alas. Another who writes in terms of common ways of knowing or of knowing on more than one level at once is W.H. Thorpe, whose *Science, Man and Morals* brings out again the interplay between science, religion, and art as ways of knowing. He accents the overall tendency toward wholeness and unity.[41] The difficulty is, of course, that tendencies can get blocked out by other social demands, such as the typical American desire to keep up with the Joneses. But what Keniston calls a deep need for psychic integration in man is often ignored in the race for success, and it's little wonder that he feels frustrated.

It is one thing to produce evidence of order, design, and unity as a counter-argument to a negative, self-destroying dualism, and the need for studying such evidence should be apparent, but it is another thing to find a specific bridge to fit everyone. We have touched on such bridges as technology itself, on history of technology, on curricula, on mental processes, on seeing, on feeling, on design, on unity, form, and common ways of knowing, but these items have to be known and added up by the interested observer. Perhaps a general philosophy would emerge from such a sum. Just to make more certain, however, I shall resort to the shady practice of leading the witness, so to speak, moving on to more general considerations of an exegetical nature and winding up with a hortatory prayer, as in the old days.

Knowing and adding up call for human intellect to function in this instance, not computers. Sooner or later the humanist will ask the scientist to draw a line, that is, if the scientist has successfully persuaded him to forsake his ivory tower. Joseph Wood Krutch, naturalist and literary critic, speaks for the concern of the humanist but with the understanding of the humanist. Among other matters, Krutch is interested in the manipulation of environment by human beings. He recently came across a book on biogeography which discussed the deliberate poisoning of a small Pacific island as an experiment to see what life came back and how. One reviewer had thought this might revitalize ecology. Krutch's comment was, "To him, I suppose, Hiroshima 'revitalized' both atomic research and the science of warfare." In deploring the pull of ecology away from its original interest in conservation to this ruthless kind of manipulation, he made the statement, "Beasts and machines are alike in that both are free from the burden of making a choice, recognizing a value, or assuming a responsibility." The moral is obvious: don't be a beast; stand up and be counted. Krutch is an affirmer who wants people to be more than conscious automata. If scientist and humanist are to get together, *both* will have to agree with the letter and spirit of this:

> I am willing [writes Krutch] to consider the possibility that we might be saved from our present perilous state by philosophy, religion, sensibility, or indeed anything that depends upon the free functioning of the human intellect and spirit. But I do not believe that we can be saved by propaganda, manipulation, and conditioning. The most complete and successful application of such methods could do no more than transform us into well-behaved puppets. And that would be, not salvation, but the damnation of an eternal death.[42]

It was this that Snow had in mind when he talked of the danger inherent in allowing the two cultures to grow farther apart.

Philosophy may in the long run be able to succeed if and where other possible panaceas fail, or philosophy in combination with one or another link mentioned so far. I don't mean the most formal philosophy promulgated by professionals or academics alone at a safe distance, but rather thinking about choices, values, and responsibilities on the part of scientists, engineers, technologists, writers, artists, sociologists, and laymen. The well-known California architect William Pereira explored this view in depth in a talk delivered in April, 1967. He was, of course, showing the tragedy of a continuing cultural split:

> I believe beyond the shadow of a doubt that unless science is based on a philosophical approach to man and his relationship to the world in which we live, there can be neither absolute value nor real satisfaction in scientific discovery and achievement. Utilizing the enormous power of our science and technology without reference to an underlying philosophy could mean the end of civilization.[43]

I must interpolate again the qualification that a "philosophy" does not mean the use of and reliance on reason exclusively, but something more akin to what Seaborg and Rabi have asked for, wisdom; and wisdom will include non-verifiable truths as well as those of the mind, the life of the spirit as much as the findings of the laboratory.

Inevitably we come back to the problems of communication, written and oral (the latter assuming once more a great significance due to technology and the development of electronic devices, including portable radios and low-priced television sets). Even if we assume that differences in life-interests and temperament can be overcome, the language barrier between well-meaning scientists and, say, artists or writers, is a considerable one. Just as in international politics, communication calls for translators and interpreters, understanding between two cultures calls for resolution of semantic difficulties via articles, conversation, and studies like this one. The same words do not even mean or suggest the same thing characteristically as they are handled or thought of every day by members of the two cultures. It is doubly fitting that since we are talking "philosophy" we let a philosopher present this predicament, but also his hope for a coming together through language and ideas, a few common ground rules, a vocabulary—in short, through translation and interpretation. I call upon Professor Will Jones of the Pomona College faculty, author of *The Sciences and the Humanities.*[44] In a passage on the different ways in which a poet and a scientist handle words, he turns to the common word "water":

> At the commonsense level, 'water' designates the colorless liquid that falls from the sky in rain, that bubbles up from the earth in springs, that surrounds us in oceans, that is used for drinking, washing, and so forth. Science characterizes 'water' in the same *kinds* of ways, though more precisely; thus 'water' is the liquid that freezes at 0° Centigrade, that boils at 100° Centigrade at sea level, that is used as the standard for measuring the specific gravity of other materials, and so on. But since there are many colorless liquids that look like water (e.g., alcohol, sap in trees) which do not have these scientific properties, and since the term 'water' has a number of extraneous emotive connotations, 'H_2O' is typically substituted for 'water' in the scientific languages.[45]

Jones then goes on to show how the humanities refine common sense. Keats may use "water" to suggest transitoriness ("Here lies one whose name was writ in water"); Shakespeare to suggest dissolution, washing, anointing, as his need dictates; Eliot to link physical thirst with spiritual purification. This single example (at the risk of appearing too simplistic) will serve to illustrate a principle in communication between the two cultures. From this point on, Jones gives other examples and then turns to the similarities between the scientific and artistic approaches to life which we have already encountered. His contribution is to show that artist and scientist can learn each other's usages of words with relatively little effort–if they will. Once communication is established, the gulf dwindles, and if all goes well, polarization is replaced by a continuum:

> Thus the sciences and the humanities are to be conceived. . . as forming a continuum, ranging from sciences such as physics, to sciences such as sociology, to history and common sense, to literature and the arts. All of these languages are at once explanatory *and* expressive processes. Something is *expressed* in the sciences (e.g., a preference for discontinuities and for particle models in Newton's theory of light, in contrast to Huygens'), but expression is generally subordinated to explanation. Something is *explained* in the arts (e.g., in *Macbeth,* what ambition will do to a man), but explanation is generally subordinated to expression.[46]

Jones' final summary from the point of view of a philosopher working with the problem of language and reality is to lay to rest quite convincingly the either-or, lazy notion that there are two mutually exclusive cultures:

> Accordingly, to say–some people do–that the world that science discloses is 'more real' than the world of art and of literature, is simply to reveal a preference for the sorts of values–externality, discreteness, order, simplicity, hierarchy, for instance, and the practical and technological consequences that depend on them–that are attainable in this linguistic medium. To say, as others do, that the world of the arts is more real is simply to express a preference for the different values realizable in this medium. *A man who tries to live exclusively in one of these 'realities' (it does not matter which one), and who accomplishes this difficult feat only by downgrading the other or closing his mind to it, is simply depriving himself of ranges of experience which he might otherwise enjoy. As soon as we understand the reciprocal relation that exists between language and reality, we will see that the sciences cannot possibly contradict the humanities. Indeed, far from contradicting, they supplement one another.*[47] (italics mine)

The reader may work out of Will Jones on semantics to communion through vocabulary via tolerance of ambiguity, which is thought to be one good test area for determining the best engineers and scientists. Tolerance, as we have seen, leads to sharing, perhaps the best bridge of all.

The ultimate goal is beyond philosophy and beyond a phrase like "unitary culture." It involves religion in a non-doctrinal sense. It may even sound a bit mystical. It may never be reached. But if man's reach should exceed his grasp, it is worth a try and the attempt alone will leave the human beings involved better off than before. I like the way Teilhard de Chardin puts it in *The Phenomenon of Man:*

> The time has come to realize that an interpretation of the universe—even a positivist one—remains unsatisfying unless it covers the interior as well as the exterior of things: mind as well as matter. The true physics is that which will, one day, achieve the inclusion of man in his wholeness in a coherent picture of the world.[48]

People incapable of this grand vision can sense the general principle and go back to the student who was asked to tell how many ways one could look at a tree. It came slowly with him, like pulling a tooth or a cork, but with some manipulation, it came.

In all the new leisure time supposedly to be given us by advances in technology there should be time for brooding about these questions. Wasn't it Shaw who said that unlimited leisure was a good definition of Hell? If, as has been said, technological values are neither hostile nor benevolent to human values, if technology cannot be considered apart from societal relationships, and if the impact of technology varies with the structure of the particular society in question, does it not behoove the two cultures—and for that matter all the sub-cultures like Big Business, High Society, the Ghetto, the Middlebrows, and the Aborigines—to make modern society a relatively unified group instead of a bickering or mutually ignoring series of splinter groups? Relatively unified is not uniform; it describes a similarity or kinship in the midst of differences, like the resemblances in the old family unit, itself now badly threatened by the pains and pressures of modern life.

In a picture gallery an American may say of Monet or Matisse that the light was never like *that*! What he means is that *he* has never seen light like that in California or Connecticut. If he goes to the south of France, he will know. The next step in his vocal reactions will involve an informed hesitancy to make snap judgments, the next to see as others see. Only intelligent people who are interested in living will, in general, make this kind of effort, however, for they are dedicated to profiting by experience, not just accumulating knowledge or money. Occasionally children and truly simple folk, both uncluttered or unfettered by pose or stereotypes, will see through surface disparateness to basic earthy patterns, and exclaim, "Look, that's like–." A poet sees this way and talks in metaphors. With adult and informed and specialized humanists and scientists, intelligent people though they are, some motivation may have to be supplied, but the process of discovery, as with the imaginary American tourist above, is also one of exposure, seeing, feeling, comparing, and reflection. Relationships will become clear. But the effort must be made. And before that, the reason for making it has to be furnished. This is more than a glib or smug rationale for an intellectual United Nations. It may be the rationale for the progress or fall of civilization as we know it.

One must begin somewhere. To the growing list of approaches discussed so far, I suggest the study of anthropology or archaeology, in a formal college course or at home. In such areas of amateur investigation, linguistics, art, sociology, humanity, and technology (tools), to name a few items of cultural significance, meet. Likewise, a little reading and thought will reveal many similarities in subject, tone, and "vocabulary," even form, between modern poetry and modern painting (as Wallace Stevens has shown in an essay of that title, published by the Museum of Modern Art in New York) and between both and music. Once one is "hooked," once one walks the fringe-like borders, he is ready to leap into larger, more metaphysical considerations of what Dr. Johnson in the eighteenth century referred to as *discordia concors,* literally a concordant discord. In this evangelical digression, which is not meant to sound smug or unctuous, I may say that engineers, scientists, and technologists must learn *not* to do all the things they can do and do other things they don't do, in order to help the world and lead fuller lives themselves. This, in part, would involve uniting with humanists, artists, and concerned laymen to bring the cultures closer together. If nations are to unite for private and public good (substitute local and universal, if you prefer), and if, as President Nixon has said, we are to get together again, there is an ample and obvious opportunity for an educated elite to iron out differences and show the way. It is one thing to *condemn* contemporary absurdity, loneliness, alienation, and defeatism; it is another to accept this slap-in-the-face challenge to people who should know better to get together and *affirm.* This may sound syrupy, but the alternatives are plain to see. The poet Auden has said it better in "September 1, 1939." The scientist Bruno Rossi points to the same conclusion by implication in his own way. In discussing the absence of a theory telling why a certain meson has a certain mass and a certain mean life or why nuclear forces have a certain strength and a certain range, Rossi says we are trying to find relationships between characteristics of nuclear forces and meson properties, hoping that laws will emerge—also hoping to discover connections between nuclear and electromagnetic phenomena. He concludes:

> Thus in this field, as in all other fields of science, the ultimate aim is to discover regularities that our senses alone cannot detect, to find relations between seemingly unrelated effects, to formulate laws capable of describing wider and wider classes of natural phenomena.
>
> An intuitive feeling for the order and the simplicity underlying natural phenomena is as essential to the creative scientists as it is to the creative artist; for to discover a scientific truth is merely to reveal some new aspect of the armory of nature.[49]

Snow has warned us that "It is dangerous to have two cultures which can't or don't communicate." Although communication need not, perhaps never can be, perfect, it can be attempted and there are many ways, as we have seen. The "don't" is a matter of personal responsibility. It seems to me that Aldous Huxley sums up both the difficulty and the proper spirit for facing it when he says, in *Literature and Science,*

> Thought is crude, matter unimaginably subtle. Words are few and can only be arranged in certain conventionally fixed ways; the counterpoint of unique events is infinitely wide and their succession indefinitely long. That the purified language of science, or even the richer purified language of literature should ever be adequate to the givenness of the world and of our experience is, in the very nature of things, impossible. Cheerfully accepting the fact, let us advance together, men of letters and men of science, into the ever-expanding regions of the unknown.[50]

"Let us advance together" is really the text of this chapter and this little book, as should be apparent by now. Once each culture learns the other's language (and the irony is that representatives of each willingly spend a great part of their lives learning foreign tongues as part of their academic, artistic, or professional careers), the problem of the two cultures should appear as a matter of proportion, as James T. Shotwell has written in "Mechanism and Culture":

> What is needed in both humanists and scientists is the Hellenic sense of just proportion, so that neither thought nor machines shall become master of life. . . .(Education) cannot be purely literary or idealistic without losing touch with the spirit of the age in which we live; it cannot be purely technical and remain education.[51]

Everywhere I turn I wander in to expressions of opinion in the same vein. It makes me wonder whether it is possible for thinking people to have missed all of these, or to have failed to absorb them if they have not. Consider W. Pauli's conviction in *Science and Western Thought* that there is an explicit and implicit imperative in our compartmentalized age to eliminate such distinctions "by a combination of rational understanding and mystical experience of unity." [52]

Go to the back files of the *New York Times* for December 27, 1954 and read Oppenheimer's Columbia address, in which he spoke of the separation of science from the common heritage of civilized society and the unhappy result of this separation in its alteration of the character of the potential audience for writers and artists, which must be men and not a specialized set of experts; an intelligible interpretation of life, Oppenheimer pointed out, calls for a *community* of men. You may repeat this process of self-education with little effort, and I mention only easily available materials. And since one beautiful summary essay is not easily available, at least in America, I am going to quote a large piece of Gerald Walters' "Unity of Knowledge and Experience," identifying key references in the chapter notes:

> The organizing of experience is a subtle process interrelating knower and known. Order is continually made out of disorder by our ability to give meaning to our experiences. We do not give meaning to our experience from any position of transcendence, but in a sense discover meanings, by responding to solicitations already in our experience. We organize the world from within experience, what Heidegger calls 'being-in-the world.' The significant reality is the human reality, the intentional structuring of experience, the *world-for-me* which is central to Merleau-Ponty's philosophy of phenomenonology. The act of observation cannot divide observer and observed, whether in art or quantum physics. Nor is it possible to draw a distinction between the fundamental processes of scientific or artistic creation. Whether expressed in the terms of Cassirer as *radical metaphor*, or of Koestler as *bisociative thought,* or of Schon as the *displacement of concepts,* creative thought emerges as the intuitive bridging of areas of experience previously unconnected, what Einstein called a *kind of combinatorial play*. Neither is this a mechanical process from which emotion and commitment are absent. On the contrary emotion plays as large a part in the production of a scientific hypothesis as in the creation of a work of art, that experience of the *magnum mysterium* which was to an Einstein our primitive perception of profoundest reason and most radiant beauty. . . .

Indeed, Polanyi's analysis of the stages of scientific thought can be applied with very little shift of emphasis to any form of creative work. Whether the act is called intuition or the tacit co-efficient of a scientific theory by which it bears an experience, the creation of a new hypothesis begins with a leap of the imagination. Every interpretation of nature is based on some intuitive concept of the nature of things, and discovery rests upon the interplay of intuition and observation, the discerning of *gestalten* that are aspects of reality. The process involves a high degree of personal judgment and decision, in which the rules of the so-called method are applied as rules of art. The role of observation is simply to supply clues for the apprehension of reality–the apprehension of reality thus gained forms in its turn a clue to future observations, that is, the process underlying verification. The process has its characteristic rhythms, but most nearly, in Polanyi's view, resembles the creation of a work of art which is firmly guided by a fundamental vision of the final whole, even though that whole can be definitely conceived only in terms of its yet undiscovered particular; foreknowledge of the final whole, however unparticularized, guides conjecture with reasonable probability. The Baconian prescription for empirical research is a travesty of the real processes by which discovery is made [which recalls Pasteur's remark about accidental discoveries happening to the prepared mind].

Scientific activity is brought then to an existential unity with the traditional forms of human understanding–not, it must be stressed, to an *identity,* but as having its common origins in the desire to elicit meaningful structures out of its own areas of experience and guided by a common creative imagination. As Oppenheimer concludes, both science and the arts belong to the *village*–his symbol of the intimate, creative existence in contrast to the *highway,* that meaningless, mechanized society. They form communities of artists and scientists bound in freedom and co-operation by the common bond of humanity. Both have a responsibility to assert the meaningfulness of existence against lack of meaning. Both the man of science and the man of art live always at the edge of mystery, surrounded by it. Both as the measure of their creation have always had to do with the harmonization of what is new with what is familiar, with the balance between novelty and synthesis, with the strug-

> gle to make perpetual order in total chaos. Science and art share a common obligation to keep our minds open and to keep them deep, to keep our sense of beauty and our ability to make it, and our occasional ability to see it in places remote, strange, and unfamiliar. Not an easy task in a great, open, windy world—a rugged time of it no doubt, but now as complementary and no longer antagonistic modes of experience.[52]

Frank Lloyd Wright years ago had seen a similar relationship between art and science. In a speech before the Chicago Arts and Crafts Society on March 6, 1901, he concluded, "Upon this faith in Art as the organic heart quality of the scientific frame of things, I base a belief that we must look to the artist brain, of all brains, to grasp the significance to society of this thing we call the Machine."[53]

One must also consider the possibility that more than recognition of similarities and mutual responsibility is necessary to achieve a symbiotic relationship between the two cultures. The present young generation has taught us the possibilities of political activism, and while one might condemn some of their methods, one must admit that as they adapt their tactics to fit popular reaction, they may yet have the last word. The exploiters of technology may one day be forced to get to the heart quality Wright spoke about. For example, in an article in the *New Republic* for January 11, 1969, Arthur Mendel takes issue with the technological prophets who are scaring people to death and widening the gap in trust and understanding between themselves and the laymen, including the student generation. In his piece, called "Robots and Rebels," Mendel discusses an earlier article by Zbigniew Brzezinski in which the latter had hailed the imminent arrival of the "technetronic" age, an age in which computers will reason as well as man, help man determine the sex of his children, control intelligence with drugs, and modify personalities. Total political surveillance will appear, a technocratic dictatorship. The university will cease being a remote ivory tower, according to Brzezinski, a mere repository of irrelevant wisdom, and the intellectual community will change from a pattern largely humanist-oriented to that of experts and specialists. Mendel calls this a regression to fetishism with man existing only to feed the idol, another example of modern false-consciousness which hypostasizes technology into forces too powerful to heed soft and fuzzy human values. Brzezinski does ask for improving the quality of life for man as man, but Mendel asks, what have storage banks and information retrieval buttons to do with such questions? He points out that five centuries of developing *homo economicus* through urban, technological, and scientific progress have resulted in the growing modern phenomenon he calls the Great Refusal, which emerges from an affluent society and can be detected in such activism as civil rights, Peace Corps work, anti-war demonstrations, and the current career choices which are turning away from engineering, law, and the military. Mendel feels that the college rebels belong to a new era in which affluence allows *society as a whole* for the first time in history to regain freedom and honesty as seen in the rise of personalism in the social sciences; humanistic psychology; the upsurge of interest in anthropology, mythology, and art; sexual freedom; freer religion; in short, a quiet revolution gradually changing the quality of our thought and experience. If Mr. B. is right and Mr. M. wrong, manufacturers and experts have a dreadful responsibility if they permit computers to determine man, the cleavage between the technical power men and the rest of us will spread, and the individual will indeed become a number without identity or living role; if B. is wrong and M. right, the technetronic boys had better build up a better lobby of press relations and showcase their products more convincingly. The Great Refusal may one day be more effective than any local housewife grape boycott. It just might be strong enough to force the two sides to understand each other and work together. We would have one culture or none.

The humanist also has a responsibility and must come half way. If he dislikes the methods of the student rebels, he must try his own brand of activism, but act he must if he is to survive. He must assimilate the Machine, as Mumford has pointed out, before he can go his own way:

> Our capacity to go beyond the machine rests upon our power to assimilate the machine. Until we have absorbed the lessons of objectivity, impersonality, neutrality, the lessons of the mechanical realm, we cannot go further in our development toward the more richly organic, the more profoundly human.[54]

Mumford's challenging thesis is that if man "treats the powerful automaton of his as a challenge to his own development" a greater synthesis may materialize in which the Machine enlarges "the provinces of culture" and "will carry an antidote to its own poison."[55] Such a synthesis is within the grasp of the artist or humanist who wants to do something about it. Such a synthesis suggests one culture, not two. If he does not actively participate in any of the ways suggested so far in this chapter, the humanist will learn the bitter truth set out by Dubos, namely, that "scientific knowledge, although it enables man to manipulate his environment, paradoxically leaves him an outsider in the world he is creating."[56] Dubos, like Mumford, hurls a challenge at both cultures and in doing so, clearly shows the need for unity:

> Indeed, a truly human concept of technology might well constitute the force which will make science once more part of the universal human discourse, because technology at its highest level must integrate knowledge of the external world and of man's nature.[57]

The rationale for the kind of activity which both Mumford and Dubos have called for, if a rationale be needed, is further seen in the words of the poet John Wain, whom I call upon again: "To turn your back on the mess and walk in the woods is a way out—for a time. But if we do not make some attempt to manage the world, it will run mad and overwhelm everything, including the woods, with its madness."[58]

We have come a long way from Hiroshima. Today one can read in the papers about the Soviet deployment of ICBM's with one thousand times the yield of the Bomb that fell on August 6, 1945. A decade has passed since C.P. Snow delivered his now-famous lecture, and the phrase the "two cultures" is still alive and kicking. This is the period when quite naturally, it seems, the young from four to eight years of age are given books with titles like this example, "The Little Red Computer."[59] "Closing the gap between our cultures is a necessity in the most abstract intellectual sense, as well as in the most practical," wrote Snow in a sentence chosen as epigraph for this final chapter. The necessity is greater than ever. In this book, I have rehearsed and updated the history and criticism of Snow's best-known work; discussed at length the interplay between technology, literature, and art since Hiroshima as an example of continuing confrontation; described contemporary theory and practice in the field of humanities education in engineering schools as illustration of what can be done in gap-bridging; and, finally, have outlined many possible approaches to abstract and practical realization of Snow's hope for rapprochement, with reasons for trying them. Facing the paradox of poor communication in an age of communications, I have tried to show the terrible responsibility of all groups who are intelligent, alerted, articulate, dedicated, active men and women of good will, to override petty differences and selfish holds on local sovereignty in an all-out effort to unite in saving modern civilization. The old motto "United we stand, divided we fall" has never had more meaning than it has at this moment in history.

A recent headline, "Computer as Aid to Control of Medieval Bibliography," made it clear that the twain can meet through controlled and well-directed technology, with man's purpose first and Machine as servant, not master. The physicist Rabi has called for wisdom. It seems to me the beginning of wisdom to note where we share, not emphasize cleavages and shrug the matter off. We grow from the former, we grow apart from the latter. The only direction that makes sense in these dangerous and exciting days is marked by many relatively new signs, among them the blending of interests and disciplines as seen in the establishment of a Society for the History of Technology, which issues a quarterly called *Technology and Culture* (see Chapter III), the proliferation of departments of the history of science in leading universities, the continuing IBM-supported Program for Technology and Society at Harvard, and the setting up of combination professorships in bio-medical engineering, sociology of science, and psychology and engineering. The notion of convergence rather than cleavage, nicely stated in 1920 as the first line of a sonnet by Edna St. Vincent Millay—"Euclid alone has looked on Beauty bare"—makes more sense today than it did fifty years ago, even though Euclidean geometry may now be old hat.

It would seem to me that all of us are men first, women first, and then humanists or social scientists or engineers or scientists, and then, perhaps, in the same order, drama specialists, historians, systems experts, or astrophysicists. One cannot help feeling that educated people and those in the process of acquiring an education have only one alternative to the hot-line, last-gasp call: to see with new eyes and seek new wisdom. The United Nations, weak as it may be, has lasted longer and accomplished more than many thought it would, and must get more strength through increasing understanding, for it still remains our best hope against destruction by the technological power of the ICBM's, which are, ironically, quite capable of rendering the world into one culture in a few minutes, a universal dead culture that would make the fall of Greece or Rome look like a mere puppet show. The United Nations needs the help of education and communication on a grand scale to speed the removal of ancient prejudices, and the help of modern technology to end starvation and suffering in many quarters of the globe. Such universal understanding may still be over the horizon, but it can now be seen to be possible at least. The fate of international cooperation depends in great part on cooperation among those local forces, political and non-political, which must eventually get together. There is no room or reason, for example, for scientists and humanists, engineers and laymen, artists and technologists to wander off on separate paths—not any more. They cannot afford to in the face of contemporary happenings. Since the members of the two cultures have various concerns with such things as missiles, on the one hand, and building a good life for decent people, on the other, how can they live with themselves if they do otherwise than mend the broken bridges and build new ones, consciously working toward understanding and mutual reward by working as a team? Their decision, and it is crucial, will cost no millions or billions—just a bit of interest and study—and will require only another Great Refusal, a refusal to repeat supinely the apathetic remark that the two cultures are far apart and will get farther apart. We have seen, I hope, in the foregoing pages, that this remark need not be true, is not true in many areas, and should be ultimately out of fashion. The great problem of our era is to enjoy the fruits of specialization without the dominance of specialists, to restore the whole man with a sense of belonging to something instead of being a comfortable automaton or an alienated zombie. It cannot be solved by any one sub-group, but it can be solved if the party lines and picket lines are crossed and one culture of interlocking special interests is established on the principles set forth in the statements of the wise and gifted men and women called upon in these pages. If their vision be true, and I believe it is, the world may yet recognize the force of the old proverb, "The many fail; the one succeeds."

Notes

Chapter I

[1] Wylie Sypher, *Literature and Technology,* Random House, New York, 1968, xv, 5.

[2] Franklin L. Ford, A program for science in Harvard College, Cambridge, Mass., February 1969, p. 3.

[3] The Alicia Patterson Fund Newsletter, June 10, 1968.

[4] Scientist and man of letters, *Yale Review,* n.s., **31**, no. 2, December 1941, 279.

[5] Bath (England) Technological University Catalog, 1968, p. 5.

[6] *The Two Cultures,* and *A Second Look,* Mentor Books, New York, 1963.

[7] *Ibid.*, p. 48.

[8] Quoted in D. Cornelius and E. St. Vincent, Eds., *Cultures in Conflict,* Scott, Foresman, Chicago, 1964, p. 23.

[9] *Ibid.*

[10] *Sciences and Humanities,* University of California Press, Berkeley, 1965. Reprinted by permission of The Regents of the University of California.

[11] See Levine and Thomas, *Scientist vs. Humanist,* Norton, New York, 1963, for basic materials on Arnold and Huxley, and the two cultures background.

[12] Varieties of literary utopias, in F. E. Manuel, Ed., *Utopias and Utopian Thought,* Houghton Mifflin, Boston, 1966, pp. 31-2.

[13] The scientist and society, *Imperial Oil Review,* December 1963, pp. 20-2.

[14] Sypher, *op. cit.*

[15] *Op. cit.*, p. 21.

[16] January-February, 1968, **56**, 28-47.

[17] *Saturday Review,* June 6, 1964.

[18] W.H. Davenport and D. Rosenthal, Eds., *Engineering: Its Role and Function in Human Society,* Pergamon Press, Elmsford, N.Y., 1967, p. 148.

[19] Paraphrased and quoted from *Science,* April 22, 1960, **131**, no. 3408, 1187-93.

[20] Personal interview, Cambridge, Mass., April 1969.

[21] Automation and imagination, *Harper's* **231**, October 1965, 96.

[22] *Science and the Shabby Curate of Poetry,* Longmans, London, 1964.

[23] Davenport and Rosenthal, *op. cit.*, p. 4.

[24] *A Runaway World?* B.B.C., London, 1968, p. 10.

[25] Pp. 2-3.

[26] *The American Language,* Knopf, New York, 1941, pp. 289-91.

[27] Adapted from Davenport and Rosenthal, *op. cit.*, introduction to Part II.

[28] Bernard Barber and Walter Hirsch, Eds., *The Sociology of Science,* Free Press, New York, 1962, p. 250. Copyright © 1962 by the Free Press, a Division of the Macmillan Company.

[29] Harvard University Press, Cambridge, Mass., 1969.

[30] See Note 19.

[31] October 4, 1967.

[32] *Op. cit.*, p. 22.

[33] *Ibid.*, p. 20.

[34] October 10, 1964 and June 1955, respectively.

[35] Harper & Row, New York, p. 227.

[36] Lapp, *op. cit.*, p. 228.

[37] Viking Press, New York, 1966.

[38] *Saturday Review,* December 2, 1967. Means, p. 15. Dubos, p. 70.

[39] Quoted in part and paraphrased from *Bulletin of Atomic Scientists,* November 1967.

Chapter II

[1] Irving Howe, *The Idea of the Modern,* Horizon Press, New York, 1967, pp. 38-9. What follows is part quotation, part paraphrase.

[2] *New Republic,* October 23, 1965, p. 22.

[3] The avant-garde: Which way is forward? *Nation,* November 18, 1961, pp. 396-7.

[4] Nihilism in contemporary literature, *Nineteenth Century* **144**, October 1948, 215, 222.

[5] Science and literature, *Commonweal,* May 13, 1966, pp. 218-21.

[6] *Fortune,* February 1966, p. 112 ff.

[7] Lord Brain, quoting P. Stubbs, *New Scientist* **24**, 448, in *Science and Man,* New York, 1966, 90.

[8] Oxford University Press, New York, 1964.

[9] Marx, p. 350.

[10] *Ibid.*, p. 364.

[11] *Ibid.*, p. 363.

[12] In *The Fitness of Man's Environment*, Smithsonian Annual II, Washington, D.C., 1968.

[13] Harvard University Press, Cambridge, Mass.

[14] Sussman, p. 156.

[15] *Ibid.*, p. 233.

[16] Vanderbilt University Press, Nashville.

[17] West, p. 134.

[18] Random House, New York.

[19] Sypher, p. xviii.

[20] *Ibid.*, p. 6.

[21] *Ibid.*. p. 175.

[22] *Ibid.*, p. 180.

[23] *Ibid.*, p. 150.

[24] Warburg, pp. 137-8, 112, 121.

[25] Oxford University Press, New York, 1950.

[26] University of North Carolina Press, Chapel Hill, 1961. Translated by M.B. Friedman.

[27] Ginestier, p. 15.

[28] *Ibid.*, p. 43.

[29] *Ibid.*, p. 156.

[30] *Ibid.*, p. 174.

[31] Grove Press, New York, 1960.

[32] Penguin Books, Baltimore, Md., 1962.

[33] McGraw-Hill, 1963.

[34] Page references for quotations from Brinnin and Read: 188, 194, 215, 246, 254.

[35] *Collected Poems,* Liveright, New York, 1916, p. 177.

[36] Scribner's, New York, 1964, p. 30.

[37] Walsh, *op. cit.*, p. 235.

[38] Oxford University Press, New York, 1967, paper.

[39] Rosenthal, *op. cit.*, p. 5.

[40] *Ibid.*, p. 10.

[41] *Ibid.*, p. 228.

[42] *Ibid.*, p. 331.

[43] Vol. 37, Autumn 1968, 642-3.

[44] *Anthology of Modern Poetry,* Hutchinson, London, 1963.

[45] Wain, Ed., *op. cit.,* pp. 31, 34

[46] Macmillan, London, 1961, pp. 27-9.

[47] A way to say what a man can see, *Saturday Review,* February 13, 1965, pp. 46-8.

[48] Peter Owen, London, 1964.

[49] *Voices,* pp. 36-7.

[50] Cambridge University Press, New York, 1966, pp. 65-6.

[51] Bowra, p. 118.

[52] The vocation of the poet in the modern world, *Poetry* **78**, July 1951, 223-32.

[53] The poet and the press, *Atlantic* **203**, no. 3, March 1959, 40-6.

[54] MacLeish, *op. cit.*

[55] *Saturday Review,* January 30, 1965, pp. 16-18.

[56] *Journal of the History of Ideas,* X, no. 1, January 1949, 88-103.

[57] *American Quarterly,* vol. 1, no. 2, Summer 1949, 126. Copyright, 1949, Trustees of the University of Pennsylvania.

[58] In Science as a humanistic discipline, *Bulletin of Atomic Scientists,* October 1968, p. 25.

[59] *Technology and the Academics,* Macmillan, New York, 1959.

[60] 1918, 1921, 1921, respectively.

[61] Public and private problems in modern drama, *Tulane Drama Review,* vol. 1, no. 3, June 1957, 58 ff.

[62] *Ibid.,* p. 68.

[63] Quoted in Mark Hillegas, *The Future as Nightmare,* Oxford University Press, London, 1967, p. 97.

[64] *Ibid.,* p. 96.

[65] Anchor Books, 1965.

[66] Letter, December 2, 1968.

[67] Theatre of the absurd, *Theatre Arts,* **46**, November 1962, 20 ff.

[68] The literary mind, *Nation* **201**, September 20, 1965, 203-6. Ref., 205-06.

[69] Letter, November 21, 1968. *Avant-Garde,* published by University of California Press, 1966.

[70] *Tulane Drama Review,* vol. 4, no. 4, May 1960, p. 6.

[71] Quoted in James Gindin, *Postwar British Fiction,* University of California Press, 1963, p. 56. Reprinted by permission of The Regents of the University of California.

[72] *Drama in a World of Science,* University of Toronto Press, 1962, pp. 47-8.

[73] *Ibid.,* p. 51.

[74] *Ibid.,* p. 55.

[75] Reflections on the theater, *American Scholar,* Winter 1967, **37**, 111-20.

[76] Evolution of the tragic hero, *Carleton Drama Review,* vol. 1, no. 1, 1955-6, 20.

[77] *Tulane Drama Review,* vol. 1, no. 3, June 1957.

[78] The theatre of Harold Pinter, *Tulane Drama Review,* vol. 6, no. 3, March 1962, 43, 54.

[79] The new English realism, *Tulane Drama Review,* vol. 7, no. 2, Winter 1962, 184.

[80] New York University Press, p. 27.

[81] Cresset Press, London.

[82] Laski, pp. 51-2.

[83] Mercury Books, London, 1963.

[84] *Ibid.,* p. 303.

[85] *New English Dramatists,* E. M. Browne, Ed., Penguin Books, Baltimore, Md., p. 15.

[86] *Ibid.,* quotations on pp. 19, 64, 75.

[87] *Tulane Drama Review,* vol. 7, no. 4, Summer 1963, 22.

[88] September 1968.

[89] *American Scholar,* Spring 1965, pp. 239-53.

[90] *The Reporter* **34**, no. 9, May 5, 1966, 32-5.

[91] Summer 1947, n.s. 36, no. 4, 604-10.

[92] *Ibid.,* p. 606.

[93] Is the novel done for? *Harper's* **186**, December 1942, 76-83.

[94] London, 1957, pp. 311-12.

[95] Scribner's, New York, 1968.

[96] John Braine, People kill people, in *Voices from the Crowd, Against the H Bomb,* p. 181. Quoted with the permission of Peter Owen Ltd., London.

[97] In *Science and the Shabby Curate of Poetry,* Longmans, London, 1964, pp. 18-19.

[98] *New Republic* **146**, April 23, 1962, 13-16.

[99] *Ibid.*, p. 15.

[100] Oxford University Press, New York, 1967, pp. 158-9.

[101] Hillegas, *op. cit.*, p. 173.

[102] Norton, New York, p. 48.

[103] *Ibid.*, p. 128.

[104] University of California Press, Berkeley, paper, 1963.

[105] Gindin, p. 27.

[106] *Ibid.*, p. 237.

[107] *So Human an Animal*, Scribner's, New York, 1968.

[108] Reprinted by permission of the publisher, Horizon Press, from *The Idea of the Modern* by Irving Howe, Copyright © 1967, p. 49.

[109] Letter, January 2, 1969.

[110] See Note 12.

[111] *New York Times*, October 22, 1968.

[112] *Op. cit.*, p. 158, italics mine.

[113] Quoted in review by Ashley Montagu, *Saturday Review*, December 14, 1968, p. 38.

[114] October 1968, p. 38. Reprinted by permission of Science and Public Affairs *(Bulletin of the Atomic Scientists)*. Copyright © 1969 by the Educational Foundation for Nuclear Science.

[115] Random House, New York, 1967.

[116] *American Scholar*, Spring 1965, pp. 257-72.

[117] K.G. Pontus Hultén, *The Machine*, Museum of Modern Art, New York, 1968, p. 13.

[118] *Ibid.*

[119] *Ibid.*

[120] Utopia, the city and the machine. Reprinted by permission from *Daedalus*, Journal of the American Academy of Arts and Sciences, Boston, Massachusetts, Volume 94, Number 2, Spring 1965, 290-1.

[121] The poem as defense, *American Scholar* **37**, Winter 1967, 85-93.

[122] Utopia and the good life, *Daedalus*, Spring 1965, pp. 454-5.

[123] Siegfried Giedion, *Space, Time and Architecture*, Harvard University Press, Cambridge, Mass., 5th ed., 1966. Gyorgy Kepes, *The New Landscape*, Paul Theobald, Chicago, 1964.

[124] *Saturday Evening Post* **232**, September 26, 1959, 103-06.

[125] One way to spell man, *Saturday Review*, May 24, 1958, pp. 8-10.

[126] Science, art, and technology, *Kenyon Review*, Spring 1939.

[127] Integral science and atomized art, *Bulletin of Atomic Scientists* **15**, no. 2, February 1959, 67. Reprinted by permission of Science and Public Affairs *(Bulletin of Atomic Scientists).* Copyright © 1969 by the Educational Foundation for Nuclear Science.

[128] January-February 1968, **56**, 28-47.

[129] Pp. 19-21.

[130] November 25, 1968, p. 69.

[131] December 3, 1968, p. 28.

[132] Hultén, *loc. cit.*

[133] P. 2

[134] *Science,* vol. 117, no. 3042, April 17, 1953, 398.

[135] The computer and the poet, *Saturday Review,* July 23, 1966 (Copyright 1966,Saturday Review, Inc.).

[136] *Saturday Review* **46**, December 7, 1963, 87-8.

[137] See Note 26.

Chapter III

[1] *Liberal Learning for the Engineer,* American Society for Engineering Education, Washington, D.C., 1968.

[2] In *Science and the Shabby Curate of Poetry,* Longmans, London, 1964, p. 101.

[3] Quoted in Sir Herbert Read, Art and life, *Saturday Evening Post* **232**, September 26, 1959, 106.

[4] The great American frustration, *Saturday Review,* July 13, 1968, pp. 13-16.

[5] University of Georgia at Athens, First Annual Symposium on Automation and Society, February 17, 1969.

[6] Kenneth Keniston, *The Uncommitted,* Harcourt, Brace & World, New York, 1965, p. 378.

[7] Quoted in T.R. West, *Flesh of Steel,* Vanderbilt University Press, Nashville, 1967, pp. 106-07.

[8] W.H. Davenport and D. Rosenthal, *Engineering: Its Role and Function in Human Society,* Pergamon Press, Elmsford, N.Y. 1967, pp. 148-9.

[9] C.C. Perrucci and W.K. LeBold, The engineer and the scientist: Student, professional, citizen , Purdue University *Bulletin* no. 125, January 1967.

[10] Davenport and Rosenthal, *op. cit.*, pp. 193-4, abridged.

[11] Harper & Row, New York, 1959, pp. 30-2.

[12] "West" Magazine, *Los Angeles Times*, March 10, 1968, pp. 16-17, offers factual and visual evidence of team work and team learning coming out of joint ventures involving engineers and sculptors in metal. See also Chapter II, section 7.

[13] Cambridge University Press, New York, 1967.

[14] Quoted in Levine and Thomas, Eds., *Scientist vs. Humanist*, Norton, New York, 1963, p. 176.

[15] Machines, technology, and the life of the mind, reprinted in Obler and Estrin, Eds., *The New Scientist*, Doubleday Anchor, New York, 1962, pp. 55, 61.

[16] *Saturday Review*, May 2, 1964, p. 51.

[17] Cited in Gerald Holton, Modern science and the intellectual tradition, *Science* **131**, no. 3408, April 22, 1960.

[18] Robert Gomer, The tyranny of progress, *Bulletin of Atomic Scientists*, February 1968.

[19] *Saturday Review*, March 2, 1968, p. 53.

[20] *Ibid.*, p. 54.

[21] *Saturday Review*, July 1, 1967, pp. 40-1.

[22] *Nation* **200**, January 4, 1965, pp. 3-5.

[23] ASEE Report, Liberal learning for the engineer, pp. 20-1.

[24] January 1969, pp. 1-19; references from pp. 4, 6-7, 15-16.

[25] From the minutes of a meeting held March 15, 1967, in the National Geographic Society Building, Washington, and from earlier discussions in the Dartmouth School of Engineering.

[26] Unpublished report in ditto form, Harvey Mudd College, Claremont, Calif., September 15, 1968.

[27] Introduction to Science and the human condition, *Bulletin of Atomic Scientists*, October 1968, p. 25.

Chapter IV

[1] Quoted in Gordon Hawkins, God and the Mafia, *The Public Interest*, no. 14, Winter 1969, p. 31.

[2] From *The Uncommitted*, Copyright © 1962, 1965, by Kenneth Keniston. Reprinted by permission of Harcourt, Brace & World, Inc., p. 441.

[3] *Ibid.*, p. 447.

[4] *The Broken Image: Man, Science, and Society*, Doubleday (Anchor), New York, 1966, p. vi.

[5] Davenport and Rosenthal, Eds., *Engineering: Its Role and Function in Human Society,* Pergamon Press, Elmsford, N.Y., 1966, p. 149.

[6] *Ibid.,* p. 150.

[7] Machines, technology, and the life of the mind, in F.C. Obler and H. Estrin, Eds., *The New Scientist,* Doubleday (Anchor), New York, 1962, p. 55.

[8] Davenport and Rosenthal, *op. cit.,* p. 43.

[9] Quoted in G. Levine and O. Thomas, Eds., *Scientist vs. Humanist,* Norton, New York, 1963, pp. 166-7.

[10] In *Daedalus,* Winter 1958, pp. 85-93; this selection reprinted in Levine and Thomas, *op. cit.,* p. 176. Reprinted by permission from *Daedalus,* Journal of the American Academy of Arts and Sciences, Boston, Massachusetts, Volume 87, Number 1.

[11] From a paper, Technics and the Future, given at the 100th anniversary of the American Association for the Advancement of Science, 1948; see also *In the Name of Sanity* by the same author, pp. 59-60.

[12] *Saturday Review,* February 15, 1969, p. 40.

[13] Fromm and Montagu quoted from *Saturday Review,* December 14, 1968, p. 38.

[14] In the Alicia Patterson Fund Newsletter, June 10, 1968, p. 9.

[15] *Architectural Record,* May 1965, p. 148; reprinted from *Daedalus,* Winter 1965; also in G. Holton, Ed., *Science and Culture,* Beacon Press, Boston, 1967.

[16] University of California Press, Berkeley, 1961.

[17] Quoted in Cornelius and St. Vincent, *Cultures in Conflict,* p. 52; reprinted from Science, literature, and culture, *Commentary,* June 1962.

[18] *Reconstruction in Philosophy,* Tokyo, 1919, quoted in Kepes, *op. cit.,* p. 28; reissued by Beacon Press, Boston, 1948.

[19] *Outline of Science,* vol. 4, 1922, quoted in Kepes, *op. cit.,* p. 25 from work published by Putnam's, New York, 1922.

[20] Issue of March 13, 1964, p. 373.

[21] Harbinger Books, Harcourt, Brace & World, New York, 1963, pp. 333-5.

[22] *Ibid.,* p. 335.

[23] Paul Theobald, Chicago, 1967. See also G. Bachelard, *Poetics of Space,* 1964.

[24] René Dubos, Science and man's nature, in G. Holton, Ed., *Science and Culture,* Boston, 1967, p. 267.

[25] Harvard University Press, Cambridge, Mass., 5th edition, 1966.

[26] Giedion, *op. cit.*, p. 562.

[27] Keniston, *op. cit.*, p. 254, italics his.

[28] *Ibid.*, p. 254.

[29] *Ibid.*, p. 256.

[30] *Ibid.*, p. 259.

[31] *New York Times*, December 16, 1968.

[32] *The Revolution of Hope*, Bantam Books, New York, 1968, p. 44.

[33] See Note 11.

[34] The poet in the machine age, *Journal of History of Ideas*, X, no. 1, January 1949, p. 103.

[35] Lord Brain, quoting Dr. Peter Stubbs from *New Scientist* **24**, 1964, p. 448, in his *Science and Man*, Chapter IV, Elsevier, New York, 1966.

[36] Brain, *op. cit.*, pp. 103-4.

[37] Phi Beta Kappa *Key Reporter*, Spring 1967, p. 3.

[38] The end of the age of separation, *Saturday Review*, May 18, 1968, pp. 23-5.

[39] *Ibid.*, p. 23. See also Arthur Koestler, *The Ghost in the Machine*, Chapter 14, Macmillan, New York, 1967.

[40] From *The Idea of a University*, quoted in Gerald Walters, *op. cit.*, pp. 7-8.

[41] Cornell University Press, Ithaca, N. Y., 1965.

[42] If you don't mind my saying so. Reprinted from *The American Scholar*, Volume 37, Number 4, Autumn 1968, pp. 572, 576, 577. Copyright © 1968 by the United Chapters of Phi Beta Kappa. By permission of the publishers.

[43] From an address entitled "The New World," given before the Founding Friends of Harvey Mudd College, Los Angeles Music Center, April 18, 1967.

[44] University of California Press, Berkeley, 1965. Reprinted by permission of The Regents of the University of California.

[45] *Ibid.*, pp. 201-02.

[46] *Ibid.*, p. 222.

[47] *Ibid.*, p. 223.

[48] Harper & Row (Torchbooks) 1961, pp. 35-6.

[49] The esthetic motivation of science, quoted in Kepes, *op. cit.*, p. 67.

[50] Quoted in Davenport and Rosenthal, *op cit.*, p. 279.

[51] Quoted in Cornelius and St. Vincent, *op cit.*, p. 130.

[52] Quoted in *Technology and Society,* vol. 4, no. 2, Gerald Walters, Unity of knowledge and experience, Bath University of Technology, 1967, p. 44. Works cited: R. Merleau-Ponty, *The Phenomenonology of Perception;* E. Cassirer, *Essay on Man;* A. Koestler, *The Act of Creation;* D.A. Schon, *Displacement of Concepts;* R. Michelmore, *Einstein;* M. Polanyi, *Science, Faith and Society.*

[53] Quoted in Kepes, *op. cit.,* p. 25.

[54] *Technics and Civilization,* Harcourt, Brace & World, New York 1963, p. 363.

[55] *Ibid.,* p. 325.

[56] Reprinted by permission from *Daedalus,* Journal of the American Academy of Arts and Sciences, Boston, Massachusetts, Volume 94, Number 1.

[57] *Ibid.*

[58] Review of Norman Mailer's *Cannibals and Christians* in *New Republic* **155**, October 1, 1966, p. 20.

[59] By Ralph Steadman, McGraw-Hill Junior Books, New York, 1969.

Selected Bibliography

Readings in Relationships
between
Science, Technology, and Engineering
and
Humanities, Social Sciences, and Fine Arts

Adams, R. McC., ed. *Fitness of Man's Environment,* Smithsonian (Random House), Washington, D.C., 1968.

Anon. Two cultures in engineering design, *Engineering,* March 13, 1964.

Arnold, M. Literature and science, in F. Mulhauser, ed., *Selected Poetry and Prose of Matthew Arnold,* Rinehart, New York, 1958.

Ashby, Sir Eric *Technology and the Academics,* Macmillan, New York, 1959.

Bachelard, G. *Poetics of Space,* Orion Press, New York, 1964.

Barber, B. *Science and the Social Order,* Collier, New York, 1962.

Barber, B. and W. Hirsch, eds. *Sociology of Science,* Free Press, New York, 1967.

Barzun, J. *Science: Glorious Entertainment,* Harper & Row, New York, 1964.

Bibby, C. Science: Tool of culture, *Saturday Review,* June 6, 1964.

Boulton, D., ed. *Voices from the Crowd (Against the H Bomb),* Peter Owen, London, 1964.

Bowra, C. M. *Poetry and Politics,* Cambridge University Press, New York, 1966.

Bowron, B., L. Marx, and A. Rose Literature and covert culture, *American Quarterly,* Winter 1957.

Bright, J. Opportunity and threat in technology, *Harvard Business Review,* November 1963.

Bronowski, J. *Science and Human Values,* Harper & Row (Torchbooks), New York, 1956.

Bronowski, J. Science as a humanistic discipline, *Bulletin of Atomic Scientists,* October 1968.

Brown, H. *Science and the Creative Spirit,* University of Toronto Press, Canada, 1958.

Buck, P. Artist in a world of science, *Saturday Review,* September 20, 1958.

Burgess, A. *The Novel Now,* Norton, New York, 1967.

Burke, J., ed. *The New Technology and Human Values,* Wadsworth, Belmont, Cal., 1966.

Burnham, J. *Beyond Modern Sculpture: Effects of Science and Technology on the Sculpture of this Century,* Braziller, New York, 1969.

Bush, D. *Science and English Poetry,* Oxford University Press, New York, 1950.

Cadden, J. and P. Brostowin, eds. *Science and Literature,* Heath (Raytheon), Boston, 1964.

Calhoun, J. Role of engineering in the university, *Engineering Education,* April 1963.

Carson, R. *Silent Spring,* Crest Books, New York, 1964.

Chase, E. Politics and technology, *Yale Review,* March 1963.

Commoner, B. *Science and Survival,* Viking Press, New York, 1966.

Conant, J. B. *Modern Science and Modern Man,* Doubleday (Anchor), New York, 1953.

Cornelius, D. and E. St. Vincent, eds. *Cultures in Conflict,* Scott, Foresman, Glenview, Ill., 1964.

Cousins, N. The computer and the poet, *Saturday Review, Inc.,* July 23, 1966.

Cross, H. *Engineers and Ivory Towers,* McGraw-Hill, New York, 1952.

Davenport, W. Resource letter on technology, literature, and art since World War II, *American Journal of Physics,* **38**: no. 4, April 1970, 407-14.

Davenport, W. and D. Rosenthal, eds. *Engineering: Its Role and Function in Human Society,* Pergamon Press, Elmsford, N.Y., 1966.

Davis, D. Art and technology—the new combine, *Art in America,* January-February 1968.

Dubos, R. *So Human an Animal,* Scribner's, New York, 1968.

Egler, F. Pesticides in our ecosystem, *American Science* **52,** 110, 1964.

Ellul, J. *The Technological Society,* Knopf, New York, 1964.

Esslin, M. Theatre of the absurd, *Tulane Drama Review,* May 1960.

Ferkiss, V. *Technological Man,* Braziller, New York, 1969.

Ferry, W.H. The Technophiliacs, *Center Magazine,* July 1968.

Fromm, E. *Revolution of Hope: Toward a Humanized Technology,* Harper & Row, New York, 1968.

Ghiselin, B. *The Creative Process,* University of California Press, Berkeley, 1952.

Giedion, S. *Space, Time and Architecture,* Harvard University Press, Cambridge, 1966.

Gindin, J. *Postwar British Fiction: New Accents and Attitudes,* University of California Press, Berkeley, 1963.

Ginestier, P. *The Poet and the Machine,* University of North Carolina Press, Chapel Hill, 1961.

Glicksberg, C. Nihilism in contemporary literature, *Nineteenth Century,* October 1948.

Goldberg, M. Impact of technological change on the humanities, *Educational Record,* Fall 1965.

Graves, R. A poet's investigation of science, *Saturday Review,* December 7, 1963.

Green, H. The new technological era: A view from the law, *Bulletin of Atomic Scientists,* November 1967.

Green, M. *Science and the Shabby Curate of Poetry,* Longmans, London, 1965.

Hauser, A. *Social History of Art,* Vintage, New York, 1951.

Hawkes, J. Automation and imagination, *Harper's,* October 1965.

Hayek, F. A. *Counter-Revolution of Science,* Free Press, New York, 1964.

Hillegas, M. *The Future as Nightmare,* Oxford University Press, New York, 1967.

Holton, G. Modern science and the intellectual tradition, *Science,* April 22, 1960.

Holton, G., ed. *Science and Culture,* Beacon Press, Boston, 1967.

Howe, I. The fiction of anti-utopia, *New Republic,* April 23, 1962.

Howe, I. *The Idea of the Modern,* Horizon Press, New York, 1967.

Hultén, K.G.P. *The Machine,* Museum of Modern Art, New York, 1968.

Huxley, A. *Literature and Science,* Harper & Row, New York, 1963.

Huxley, T. H. *Science and Culture,* D. Appleton, London, 1882.

Johnson, J. *The Inland Island,* Simon & Schuster, New York, 1969.

Jones, W. T. *The Sciences and the Humanities,* University of California Press, Berkeley, 1965.

Kazin, A. The literary mind, *Nation,* September 20, 1965.

Kazin, A. Imagination and the age, *Reporter,* May 5, 1966.

Keniston, K. *The Uncommitted,* Harcourt, Brace & World, New York, 1962, 1965.

Kepes, G. *The New Landscape in Science and Art,* Paul Theobald, Chicago, 1964.

Kepes, G. Visual arts and the sciences, *Daedalus,* Winter 1965.

Klaw, S. *The New Brahmins,* Morrow, New York, 1968.

Krutch, J. W. If you don't mind my saying so, *American Scholar,* Autumn 1968.

Lapp, R. *The New Priesthood,* Harper & Row, New York, 1965.

Levine, G. and O. Thomas, eds. *Scientist vs. Humanist,* Norton, New York, 1963.

Lifton, R. J. *Death in Life: Survivors of Hiroshima,* Random House, New York, 1967.

Lindsay, R. S. *Role of Science in Civilization,* Harper, New York, 1964.

Littlejohn, D. The anti-realists, *Daedalus,* Spring 1963.

Loney, G. Theatre of the absurd, *Theatre Arts,* November 1962.

MacLeish, A. Poet and the press, *Atlantic,* March 1959.

MacLeish, A. The great American frustration, *Saturday Review,* July 13, 1968.

MacShane, F. The new poetry, *American Scholar,* Autumn 1968.

Manuel, F., ed. *Utopia and Utopian Thought,* Houghton Mifflin, Boston, 1966.

Marine, G. *America the Raped,* Simon & Schuster, New York, 1969.

Marx, L. *Machine in the Garden,* Oxford University Press, New York, 1964.

Matson, F. W. *Broken Image (Man, Science and Society)* Doubleday (Anchor), New York, 1966.

Means, R. Why worry about nature? *Saturday Review,* December 6, 1967.

Mesthene, E., ed. *Technology and Social Change,* Bobbs-Merrill, Indianapolis, Ind., 1967.

Mesthene, E. How technology will shape the future, *Science,* July 12, 1968.

Morris, C. Science, art and technology, *Kenyon Review,* Spring 1939.

Muller, H. J. *Science and Criticism,* Yale University Press, New Haven, Conn., 1958.

Muller, H. Scientist and man of letters, *Yale Review,* December 1941.

Mumford, L. *Technics and Civilization,* Harbinger Books, New York, 1963.

Mumford, L. *Art and Technics,* Columbia University Press, New York, 1952.

Mumford, L. *Myth of the Machine,* Harcourt, Brace & World, New York, 1967.

Nelson, N. Science and Irresponsible Imagination, *Yale Review,* Spring 1953.

Obler, F. C. and H. Estrin, eds. *The New Scientist,* Doubleday (Anchor), New York, 1962.

Orwell, G. *Road to Wigan Pier,* Berkley, New York, 1967.

Packard, V. *Waste Makers,* Pocket Books, New York, 1960.

Parsons, J. Myths, emotions, and the great audience, *Poetry,* November 1950.

Peacock, R. Public and private problems in modern drama, *Tulane Drama Review,* March 1959.

Pearson, N. H. The American poet in relation to science, *American Quarterly,* Summer 1949.

Polanyi, M. *Science, Faith and Society,* University of Chicago Press, 1964.

Price, Derek *Little Science, Big Science,* Columbia University Press, New York, 1963.

Price, Don *Government and Science,* Oxford University Press, New York, 1962.

Prior, M. *Science and the Humanities,* Northwestern University Press, Evanston, Ill., 1962.

Pronko, L. *Avant-Garde,* University of California Press, Berkeley, 1966.

Rabi, I. Scientist and humanist, *Atlantic,* January 1956.

Rabinowitch, E. Integral science and atomized art, *Bulletin of Atomic Scientists,* February 1959.

Ramo, S. New pervasiveness of engineering, *Engineering Education,* October 1962.

Raudsepp, E. The creative engineer, *Machine Design,* May-June 1959.

Read, Sir Herbert Art and life, *Saturday Evening Post,* September 26, 1959.

Richards, I. A. *Science and Poetry,* Harcourt, Brace & World, New York, 1926.

Rosenthal, M. L. *The New Poets,* Oxford University Press, New York, 1967.

Saturday Review Science Section, December 2, 1967.

Schon, D. *Technology and Change,* Delacorte Press, New York, 1967.

Schwartz, D. Vocation of the poet in the modern world, *Poetry,* July 1951.

Sewell, E. Science and literature, *Commonweal,* May 13, 1966.

Silberman, C. E. Is technology taking over?, *Fortune,* February 1966.

Snow, C. P. *The Two Cultures* (and *A Second Look),* Mentor Books, New York, 1964.

Stegner, W. Is the novel done for?, *Harper's,* December 1942.

One way to spell man, *Saturday Review*, May 24, 1958.

Sussman, H. *Victorians and the Machine,* Harvard University Press, Cambridge, Mass., 1968.

Swenson, M. Poet as anti-specialist, *Saturday Review,* January 30, 1965.

Sypher, W. Poem as defense, *American Scholar,* Winter 1967.

Sypher, W. *Literature and Technology,* Random House, New York, 1968.

Technology and Culture *Symposium: Technology for Man.* Includes W.E. Howland, Engineering education for social leadership; J. Burke, Let's be sure technology is for man; S. Florman, Engineers and the end of innocence; J. Wallace, Engineering use of human beings. January 1969.

Thackrey, T. O. Coming struggle to breathe, *Saturday Review,* October 10, 1964.

Thorpe, W. H. *Science, Man and Morals,* Cornell University Press, Ithaca, N. Y. 1965.

Traschen, I. Modern literature and science, *College English,* January 1964.

Trilling, O. New English realism, *Tulane Drama Review,* Winter 1962.

Viereck, P. The poet in the machine age, *Journal of History of Ideas,* January 1949.

Von Neumann, J. Can we survive technology?, *Fortune,* June 1955.

Waggoner, H. H. *The Heel of Elohim,* Oklahoma University Press, Norman, 1950.

Warburg, J., ed. *The Industrial Muse,* Oxford University Press, New York, 1958.

Warburg, J. Poetry and industrialism, *Modern Language Review,* April 1958.

Weinberg, A. *Reflections on Big Science,* M.I.T. Press, Cambridge, Mass., 1967.

Wellwarth, G. *Theater of Protest and Paradox,* New York University Press, New York, 1964.

West, T. R. *Flesh of Steel,* Vanderbilt University Press, Nashville, 1967.

White, L. Discipline of history of technology, *Engineering Education,* June 1964.

White, L. *Medieval Technology and Social Change,* Oxford University Press, New York, 1962.

Wickham, G. *Drama in a World of Science,* Toronto University Press, Canada, 1962.

Wilson, J. T. On the history of science, *Saturday Review,* May 2, 1964.

Wilson, J. T. Science is everybody's business, *American Scientist* **52**, 266A, 1964.

Wind, E. Long battle between art and machine, *Harper's,* February 1964.

Woodbury, R. Scientists, engineering, and humanities, *Technology Review,* M.I.T., **63**, 147, 1959.

Index